NATEF Standards Job Sheets

Engine Repair

(Test A1)

Jack Erjavec

THOMSON

DELMAR LEARNING™

Australia   Canada   Mexico   Singapore   Spain   United Kingdom   United States

THOMSON

DELMAR LEARNING

NATEF Standards Job Sheets

Engine Repair
(Test A1)

Jack Erjavec

Business Unit Director:
Alar Elken

Executive Editor:
Sandy Clark

Acquisitions Editor:
Sanjeev Rao

Team Assistant:
Matthew Seeley

Executive Marketing Manager:
Maura Theriault

Marketing Coordinator:
Brian McGrath

Channel Manager:
Fair Huntoon

Executive Production Manager:
Mary Ellen Black

Production Manager:
Larry Main

Production Editor:
Betsy Hough

Cover Design:
Michael Egan

ISBN: 0-7668-6367-0

NOTICE TO THE READER

CONTENTS

PREFACE

With every passing day it becomes harder to learn all that it takes to become a competent automotive technician. Technological advancements have allowed automobile manufacturers to build safer, more reliable, and more efficient vehicles. This is great for consumers, but along with each advancement comes a need for more knowledge.

Fortunately, students don't need to know it all. In fact, no one person knows everything about everything in an automobile. Although they don't know everything, good technicians do have a solid base of knowledge and skills. The purpose of this book is to give students a chance to develop all the skills and gain the knowledge of a competent technician. It is also the purpose of the guidelines established by the National Automotive Technicians Education Foundation (NATEF).

At the expense of much time and the effort of many minds, NATEF has assembled a list of basic tasks for each of its certification areas. These tasks identify the basic skills and knowledge levels of competent technicians. The tasks also identify what is required for a student to start a successful career as a technician.

Most of the content in this book consists of job sheets. These job sheets relate to the tasks specified by NATEF. The main considerations in the creation of these job sheets were student learning and program certification by NATEF. Students are guided through standard industry-accepted procedures. While they are progressing, students are asked to report their findings and offer their thoughts on the steps they have just completed. The questions asked of the students are thought provoking and require students to apply what they know to what they observe.

The job sheets were designed to be generic; that is, whenever possible, the tasks can be performed on any vehicle from any manufacturer. Completion of the sheets does not require the use of specific brands of tools and equipment; instead, students use what is available. In addition, the job sheets can be used as a supplement to any good textbook.

Words to the Instructor: I suggest you grade these job sheets on completion and reasoning. Make sure the students answer all questions, and then look at the reasons to see if the task was actually completed and to get a feel for their understanding of the topic. It'll be easy for students to copy others' measurements and findings, but each student should have his or her own base of understanding, and that will be reflected in the explanations given.

Words to the Student: While completing the job sheets, you have a chance to develop the skills you need to be successful. When asked for your thoughts or opinions, think about what you observed. Think about what could have caused those results or conditions. You are not being asked to give accurate explanations for everything you do or everything you observe. You are only asked to think. Thinking leads to understanding. Good technicians are good because they have a basic understanding of what they are doing and of what they are doing it to.

NOTICE: SOME PARTS OF THIS COPY ARE DIFFERENT THAN THE PREVIOUS PRINTING OF THIS BOOK. THE CONTENTS HAVE BEEN UPDATED IN RESPONSE TO THE RECENT CHANGES MADE BY NATEF. *SEE PAGE 201 FOR DETAILS.*

ENGINES

To prepare you to learn what you should learn from completing the job sheets, some basics must be covered. This discussion begins with an overview of engines. Emphasis is placed on what they do and how they work. This includes the major components and designs of engines and their role in the efficient operation of engines of all designs.

Preparing to do something on an automobile would not be complete if certain safety issues were not addressed. A discussion of safety covers those things you should and should not do when working on engines. Included are proper ways to deal with hazardous and toxic materials.

NATEF's task list for Engine Repair certification is also given with definitions of some of the terms used to describe the tasks. This list gives you a good look at what the experts say you need to know before you can be considered competent to work on engines.

Following the task list are descriptions of the various tools and types of equipment you need to be familiar with. These are the tools you will use to complete the job sheets. They are also the tools NATEF has identified as being necessary for servicing engines.

Following the tool discussion is a cross-reference guide that shows what NATEF tasks are related to specific job sheets. In most cases there are single job sheets for each task. Some tasks are part of a procedure and when this occurs one job sheet may cover two or more tasks. The remainder of the book contains the job sheets.

BASIC ENGINE THEORY

The engine provides the power to drive the wheels of the vehicle. All gasoline and diesel automobile engines are classified as internal combustion engines because the combustion or burning that creates energy takes place inside the engine. Combustion is the burning of an air and fuel mixture. As a result of combustion, large amounts of pressure are generated in the engine. This pressure or energy is used to power the car. The engine must be built strong enough to hold the pressure and temperatures formed by combustion.

Four-Stroke Cycle

Nearly all automotive engines run through the four-stroke cycle to produce power (Figure 1). These strokes or events repeat themselves, in each cylinder, several times per minute. The name of the stroke defines what event takes place at each stroke. There is the intake stroke during which air and fuel are pushed into the cylinder. The compression stroke compresses the air and fuel mixture to prepare it for ignition. The power stroke is the one that supplies the power. It occurs after ignition and results from the quick expansion of the air/fuel mixture during combustion.

Engine Identification

Before you begin any work on an engine you need to make a positive identification of the engine. It is best to do this by locating the Vehicle Identification Number (VIN). The VIN is visible through the windshield on the driver's side of the vehicle. The coding used in the VIN can tell you much about the vehicle, including what engine the vehicle was originally manufactured with.

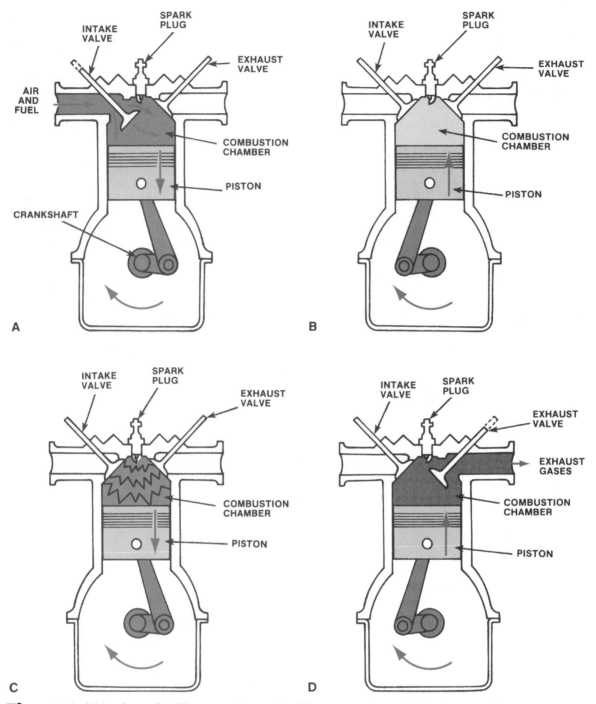

Figure 1   (A) Intake stroke, (B) compression stroke, (C) power stroke, and (D) exhaust stroke.

Cylinder Block

The biggest part of the engine is the cylinder block. The cylinder block is a large casting of metal that is bored with holes to allow for the passage of lubricants and coolant through the block and provide spaces for movement of mechanical parts. The block contains the cylinders, which are round passageways fitted with pistons. Combustion takes place in the cylinders as the pistons move through their four-stroke cycle. A cylinder block is normally one piece, cast, and machined so that all of the parts contained in it fit properly. Blocks are typically made of iron or aluminum.

The cylinder bores are critical areas of service for a technician. The walls of the bore must be the correct size and be without flaws in order to allow the piston to move smoothly up and down the bore while it maintains a seal against the walls.

Pistons

The burning of air and fuel takes place between the cylinder head and the top of the piston. The piston is a can-shaped part closely fitted inside the cylinder. In a four-stroke cycle engine, the piston moves through four different movements or strokes to complete one cycle. On the intake stroke, the piston moves downward, and a charge of air/fuel mixture is introduced into the cylinder. As the piston travels upward, the air/fuel mixture is compressed in preparation for burning. Just before the piston reaches the top of the cylinder, ignition occurs and combustion starts. The pressure of expanding gases forces the piston downward on its power stroke. When it reciprocates, or moves upward again, the piston is on the exhaust stroke. During the exhaust stroke, the piston pushes the burned gases out of the cylinder.

The gap between the outside of the piston and the cylinder walls is sealed by piston rings. These rings of steel are formed under tension so that they expand outward from the piston to the cylinder walls to form a good seal.

If the piston doesn't seal well, the efficiency of its four strokes decreases. Consider the intake stroke. The piston moves down to increase the volume of the cylinder and to lower the pressure in the cylinder. If there is leakage around the piston rings, air from the crankcase, which is below the piston, will enter the cylinder. This in turn will lower the amount of vacuum formed and reduce the amount of air/fuel mixture that will be drawn into the cylinder. The other strokes are affected in the same way.

Making sure the piston is able to move freely up and down the cylinder while maintaining a seal is a critical task during an engine overhaul.

Connecting Rods and Crankshaft

The up and down motion of the pistons must be converted to rotary motion before it can drive the wheels of a vehicle. This conversion is achieved by linking the piston to a crankshaft with a connecting rod. The upper end of the connecting rod moves with the piston. The lower end of the connecting rod is attached to the crankshaft and moves in a circle. The end of the crankshaft is connected to the transmission to continue the power flow through the drive train and to the wheels.

The crankshaft is housed by the engine block and is held in place by main bearing caps. Between the main bearing caps and the crankshaft and the cylinder block and the crankshaft are flat insert bearings.

The crankshaft does not rotate directly on the bearings; instead, it rotates on a film of oil trapped between the bearing surface and the crankshaft journals. The area that holds the oil film is critical to the life of an engine. If the crankshaft journals become out-of-round, tapered, or scored, or if the bearings are worn, the proper oil film will not form. Without the correct oil film, premature wear will occur on the bearings and crankshaft journals. This can also result in crankshaft breakage. Therefore, another critical area for technicians is the one between the crankshaft journals and the bearing surfaces.

Cylinder Head

The cylinder head mounts on top of the engine block and serves as the uppermost seal for the combustion chamber. To aid in that sealing, a head gasket is sandwiched between the cylinder head and engine block. The combustion chamber is an area into which the air/fuel mixture is compressed and burned. The cylinder head contains all or most of the combustion chamber. The cylinder head also contains ports, which are passageways through which the air/fuel mixture enters and burned gases exit the cylinder. The head also contains the valves, which must open to allow air/fuel in and allow exhaust gases out. When the valves are closed, they too need to provide a positive seal. The cylinder head may also be the mounting spot or serve as the housing for a camshaft.

Much of the work done by a technician on a cylinder head is done to ensure a good seal. A good seal is formed between the deck of the block and the cylinder head when both surfaces are flat and parallel to each other. This not only provides a seal for the combustion chamber but also allows the oil and coolant passages in the head to positively connect with passages in the engine block.

Valve Train

A valve train is a series of parts used to open and close the intake and exhaust ports. A valve is a movable part that opens and closes a passageway. A camshaft controls the movement of the valves, causing them to open and close at the proper time. Springs are used to help close the valves.

An engine's valves close into a valve seat. The interference fit between the face of the valve and the valve seat provide for the seal. Since this seal is made by two metal surfaces, it is critical that the surfaces be smooth and perfectly round.

The valves move in a valve guide that is typically pressed into the cylinder head. The clearance between the stem of the valve and the inside diameter of the guide provides some lubrication for the valve's movement. If the clearance is excessive or irregular, the valve may become canted while it moves and may not provide for a good seal when it is closed. Excessive valve to guide clearance will also allow oil to enter into the combustion chamber from the valve train. To prevent excessive oil from passing through the guides, oil seals are installed on the valve guides or valve stems.

A valve has two primary purposes: it must open and it must close. The opening of the valve is controlled directly or indirectly by a camshaft. The closing of the valve is the primary job of a valve spring. The valve spring forces the valve closed when the camshaft no longer is putting pressure on the valve.

The valve spring is held in place by a spring retainer and valve locks or keepers. This assembly also works to hold the valve in place in the cylinder head.

All current engines have at least one camshaft; some have more. Vee-type engines with dual overhead camshafts have four. Regardless of the number of camshafts an engine has, there is one camshaft lobe for each valve. That lobe is responsible for the opening and closing of that valve. The rotation of a camshaft is timed with the crankshaft, at half the speed of the crank. The camshaft and crankshaft must be perfectly in time with each other in order for the engine to efficiently pass through the four strokes of the engine's pistons.

The sides of a camshaft lobe have ramps that control how quickly the valve opens and closes. The highest point of the lobe gives the maximum valve opening. When the valve is fully open, more air/fuel can enter the cylinder or more exhaust gas can exit the cylinder. The base circle of the lobe provides no valve lift and the valve is closed.

Camshafts rotate on insert bearings, similar to those used with the crankshaft. Camshaft bearings for OHV engines and some OHC engines are full-round insert bearings. The bearings are pressed into bores in the engine block or cylinder head. Many OHC engines use split bearings. These engines have bearing caps that secure the camshaft in the cylinder head or in a camshaft housing mounted to the top of the cylinder head.

Meshed gears or a timing chain and gears can drive the camshaft in an OHV engine. The camshaft(s) of an OHC engine are either driven by a timing belt(s) and sprockets or timing chain(s) and gears. Regardless of the drive mechanisms, the gears and sprockets must be timed with the crankshaft. Always refer to the appropriate service manual when servicing camshaft drive systems.

The valve train consists of a combination of these parts: lifter, pushrod, and rocker arm. OHV engines use all three to transfer the movement of the camshaft lobes to the valves. On OHC engines, the camshaft may directly or indirectly move the valves. When the camshaft directly controls the valves, the camshaft is placed above the valves and the lobes ride on the top of the valves. Often, the camshaft indirectly drives the valves. These setups utilize a compact setup with a rocker arm and lifter.

It is important that the correct clearance between the top of the valve and the mechanism operating the valve be maintained. The gap allows for the expansion of parts caused by heat. If the gap is too small, the expanded size of the parts may cause the valves to remain slightly open when they should be closed. The advantage of hydraulic lifters is that they compensate for the expansion and do not require periodic adjustment. All mechanical lifters require valve lash adjustments. The method used to adjust valve lash varies from valve train to valve train. The procedure may be simply adjusting a screw or it may involve replacing a shim.

Intake and Exhaust Systems

A manifold is metal ductwork assembly used to direct the flow of gases to or from the combustion chambers. Two separate manifolds are attached to the cylinder head. The intake manifold delivers a mixture of air and fuel to the intake ports. The exhaust manifold mounts over the exhaust ports and carries exhaust gases away from the cylinders.

The intake and exhaust systems are critical to the efficient operation of an engine. An engine needs air to operate. This air is drawn into the cylinders through the intake or intake induction system during the intake stroke of the pistons. This air is then mixed with fuel and delivered to the combustion chambers.

After combustion, burnt gases are present in the combustion chamber. In order to make room for a fresh supply of intake air and fuel, those exhaust gases must leave the combustion chamber. The exhaust system has that responsibility.

Although there seems to be an endless supply of air outside the engine, often not enough air is available for the engine. Because the actual time of the intake stroke is so short, there is little time to gather, direct, and force air into the cylinders. A vehicle's air induction system is designed to move as much air as possible and as quietly as possible. This statement is more true for high performance vehicles than normal vehicles, simply because engine performance is directly proportional to intake airflow.

A typical intake system consists of an air cleaner assembly, air filter, air ductwork, and intake manifold. These components are designed to quickly deliver a fresh and clean supply of air to the engine's cylinders. The design of the components determines the maximum amount of air that can be delivered and the actual amount is determined by the position of the engine's throttles plates and the vacuum formed on the pistons' intake stroke.

Many engines have been equipped with turbochargers and superchargers to increase the amount and speed of air delivery to the engine. A supercharger is driven by the crankshaft with a drive belt. A turbocharger uses the flow of exhaust gases and is not mechanically connected to the engine. Both of these devices are designed to force pressurized air into the combustion chambers during their intake strokes.

A highly efficient intake system is worthless if the exhaust gases cannot leave the combustion chamber. In order for the cylinders to form a strong vacuum or low pressure, the cylinders need to be empty. The exhaust system has this responsibility. Once the exhaust has left the cylinder, a strong vacuum can be formed and a fresh supply of intake air and fuel can enter the cylinder. This allows for the beginning of another four-stroke cycle.

As the piston moves up on its exhaust stroke, the exhaust gases are pushed out into the exhaust system. From there, the gases must travel quite a distance to leave the exhaust system and be released into the atmosphere. To be able to exit the exhaust system, the exhaust gases must remain under pressure and be hotter than the outside air. As the exhaust moves through the exhaust system, it cools some and the pressure drops. The flow of exhaust will always move from a point of higher pressure to a point of lower pressure. Therefore, as long as the pressure right in front of the moving exhaust gases is lower than the pressure of the exhaust, the exhaust will continue to flow out. An efficient exhaust system ensures that this happens.

Exhaust systems normally face one of two problems: leaks or restrictions. Leaks not only cause excessive noise, but can also decrease the temperature and pressure of the exhaust. Restrictions cause exhaust pressure to build up before the restriction. The difference in pressure determines the rate of flow. When the pressure difference is slight, less exhaust will be able to leave the cylinders and there will be less room for the fresh intake charge.

Lubrication System

The moving parts of an engine need constant lubrication. Lubrication limits the amount of wear and reduces the amount of friction in the engine. Friction occurs when two objects rub against each other, generating heat. The engine relies on the circulation of engine oil to keep things healthy and to minimize wear.

Oil is delivered under pressure to all moving parts of the engine. Engine damage will occur if dirt gets trapped in small clearances designed for an oil film or if dirt blocks an oil passage. The oil circulating through the engine is filtered by an oil filter that should be replaced whenever the oil is changed. Periodic oil and filter changes ensure that clean oil circulates throughout the engine.

A screen on the oil pickup tube initially filters the engine's oil. The oil pump draws oil out of the oil pan through the pickup tube. The oil pan serves as a reservoir for the oil. Oil drawn from the pan is circulated through the engine. After circulating, the oil drips back into the oil pan.

The oil pump is driven by the crankshaft or camshaft. The pump is responsible for the circulation of pressurized oil throughout the engine. Since an oil pump is a positive displacement pump, an oil pressure relief valve is needed to prevent excessively high oil pressures. Pressure is controlled to make sure the oil will flow over the parts, not spray over the parts.

The pressurized oil travels through oil passages or galleries in the engine, crankshaft, connecting rods, and other parts.

Problems with the engine's lubrication system will cause engine problems. Sometimes lubrication problems are not obvious, and at other times (such as when there are oil leaks) they are very obvious. Likewise, sometimes the causes of these problems can be quite obvious, and at other times not so obvious. Diagnosis of the system can be challenging.

Cooling System

The burning of the air/fuel mixture in the combustion chambers of the engine produces large amounts of heat. This heat must not be allowed to build up and must be reduced, because it can easily damage and warp the metal parts of an engine. To remove this heat, a heat-absorbing liquid, called engine coolant, circulates inside the engine. The system responsible for the circulation of coolant and the dissipation of heat is called the engine's cooling system.

A typical cooling system relies on a water pump that circulates the coolant through the system. The engine typically drives the pump. The coolant is a mixture of water and antifreeze. The coolant is pushed through passages, called water jackets, in the cylinder block and head to remove heat from the area around the cylinders' combustion chambers. The heat picked up by the coolant is sent to the radiator. The radiator transfers the coolant's heat to the outside air as the coolant flows through its tubes. To help remove the heat from the coolant, a cooling fan is used to pull cool outside air through the fins of the radiator. The cooled liquid is then returned to the engine to repeat the cycle.

Since parts of the cooling system are located in various spots under the vehicle's hood, hoses are used to connect these parts and keep the system sealed.

A thermostat is used to control the temperature by controlling the amount of coolant that moves into the radiator. When the engine is cold, the thermostat is closed and no coolant flows to the radiator. This allows the engine to warm up to the correct operating temperature. As the coolant warms, the thermostat opens and allows coolant to flow through the radiator. The hotter the coolant gets, the wider the thermostat is open.

Engine coolant is actually a mixture of antifreeze/coolant and water. This mixture lowers the freezing point of the coolant and raises its boiling point. The pressure of the system also increases the boiling point of the coolant. The cooling system is a closed system. The pressure in the system is the result of the operation of the water pump and the increase in temperature. As temperature increases, so does pressure.

To raise the boiling point of the coolant even further, the cooling system is pressurized. The radiator pressure cap maintains this pressure. The cap is designed to keep the cooling system sealed until a particular pressure is reached. At that time, the cap allows some of the pressure to be vented from the system. This action prevents excessive cooling system pressures.

Not only can cooling system problems affect the durability of an engine; they can also cause many driveability problems. Today's computer controls are set to keep the engine operating at the best temperature for efficiency. This efficiency results in lower emissions and improved driveability. Problems in the cooling and lubrication systems can cause the engine to run at higher than normal temperatures; likewise some problems in the cooling system can cause the engine to run cooler than normal.

SAFETY

In an automotive repair shop, there is great potential for serious accidents, simply because of the nature of the business and the equipment used. When people are careless, the automotive repair industry can be one of the most dangerous occupations. But, the chances of your being injured when working on a car are close to nil if you learn to work safely and use common sense. Shop safety is the responsibility of everyone in the shop.

Personal Protection

Some procedures, such as grinding, result in tiny particles of metal and dust that are thrown off at very high speeds. These metal and dirt particles can easily get into your eyes, causing scratches or cuts on your eyeball. Pressurized gases and liquids escaping a ruptured hose or hose-fitting can spray a great distance. If these chemicals get into your eyes, they can cause blindness. Dirt and sharp bits of corroded metal can easily fall down into your eyes while you are working under a vehicle.

Eye protection should be worn whenever you are exposed to these risks. To be safe, you should wear safety glasses whenever you are working in the shop. Some procedures may require that you wear other eye protection in addition to safety glasses. For example, when cleaning parts with a pressurized spray, you should wear a face shield. The face shield not only gives added protection to your eyes but also protects the rest of your face.

If chemicals such as battery acid, fuel, or solvents get into your eyes, flush them continuously with clean water. Have someone call a doctor and get medical help immediately.

Your clothing should be well fitted and comfortable but made of strong material. Loose, baggy clothing can easily get caught in moving parts and machinery. Some technicians prefer to wear coveralls or shop coats to protect their personal clothing. Your work clothing should offer you some protection but should not restrict your movement.

Long hair and loose, hanging jewelry can create the same type of hazard as loose-fitting clothing. They can get caught in moving engine parts and machinery. If you have long hair, tie it back or tuck it under a cap.

Never wear rings, watches, bracelets, and neck chains. These can easily get caught in moving parts and cause serious injury.

Always wear shoes or boots of leather or similar material with non-slip soles. Steel-tipped safety shoes can give added protection to your feet. Jogging or basketball shoes, street shoes, and sandals are inappropriate in the shop.

Good hand protection is often overlooked. A scrape, cut, or burn can limit your effectiveness at work for many days. A well-fitted pair of heavy work gloves should be worn during operations such as grinding and welding or when handling hot components. Always wear approved rubber gloves when handling strong and dangerous caustic chemicals.

Many technicians wear thin, surgical-type latex gloves whenever they are working on vehicles. These offer little protection against cuts but do offer protection against disease and grease buildup under and around your fingernails. These gloves are comfortable and are quite inexpensive.

Accidents can be prevented simply by the way you act. The following are some guidelines to follow while working in a shop. This list does not include everything you should or shouldn't do; it merely presents some things to think about.

- Never smoke while working on a vehicle or while working with any machine in the shop.
- Playing around is not fun when it sends someone to the hospital. Such things as air nozzle fights, creeper races, and practical jokes have no place in the shop.
- To prevent serious burns, keep your skin away from hot metal parts such as the radiator, exhaust manifold, tailpipe, catalytic converter, and muffler.
- Always disconnect electric engine cooling fans when working around the radiator. Many of these will turn on without warning and can easily chop off a finger or hand. Make sure you reconnect the fan after you have completed your repairs.

■ When working with a hydraulic press, make sure the pressure is applied in a safe manner. It is generally wise to stand to the side when operating the press.

■ Properly store all parts and tools by putting them away in a place where people will not trip over them. This practice not only cuts down on injuries, but also reduces time wasted looking for a misplaced part or tool.

Work Area Safety

Your entire work area should be kept clean and safe. Any oil, coolant, or grease on the floor can make it slippery. To clean up oil, use a commercial oil absorbent. Keep all water off the floor. Water not only makes smooth floors slippery, but it also is dangerous as a conductor of electricity. Aisles and walkways should be kept clean and wide enough to allow easy movement. Make sure the work areas around machines are large enough so that the machinery can be safely operated.

Gasoline is a highly flammable volatile liquid. Something that is flammable catches fire and burns easily. A volatile liquid is one that vaporizes very quickly. Flammable volatile liquids are potential firebombs. Always keep gasoline or diesel fuel in an approved safety can and never use gasoline to clean your hands or tools.

Handle all solvents (or any liquids) with care to avoid spillage. Keep all solvent containers closed, except when pouring. Proper ventilation is very important in areas where volatile solvents and chemicals are used. Solvents and other combustible materials must be stored in approved and designated storage cabinets or rooms with adequate ventilation. Never light matches or smoke near flammable solvents and chemicals, including battery acids.

Oily rags should also be stored in an approved metal container. When these oily, greasy, or paint-soaked rags are left lying about or are not stored properly, they can cause spontaneous combustion. Spontaneous combustion results in a fire that starts by itself, without a match.

Disconnecting the vehicle's battery before working on the electrical system, or before welding, can prevent fires caused by a vehicle's electrical system. To disconnect the battery, remove the negative or ground cable from the battery and position it away from the battery.

Know where all of the shop's fire extinguishers are located. Fire extinguishers are clearly labeled as to what type they are and what types of fire they should be used on. Make sure you use the correct type of extinguisher for the type of fire you are dealing with. A multipurpose dry chemical fire extinguisher will put out ordinary combustibles, flammable liquids, and electrical fires. Never put water on a gasoline fire because that will just cause the fire the spread. The proper fire extinguisher will smother the flames.

During a fire, never open doors or windows unless it is absolutely necessary; the extra draft will only make the fire worse. Make sure the fire department is contacted before or during your attempt to extinguish a fire.

Tool and Equipment Safety

Careless use of simple hand tools such as wrenches, screwdrivers, and hammers causes many shop accidents that could be prevented. Keep all hand tools grease-free and in good condition. Tools that slip can cause cuts and bruises. If a tool slips and falls into a moving part, it can fly out and cause serious injury.

Use the proper tool for the job. Make sure the tool is of professional quality. Using poorly made tools or the wrong tools can damage parts or the tool itself, or could cause injury. Never use broken or damaged tools.

Safety around power tools is very important. Serious injury can result from carelessness. Always wear safety glasses when using power tools. If the tool is electrically powered, make sure it is properly grounded. Before using it, check the wiring for cracks in the insulation, as well as for bare wires. Also, when using electrical power tools, never stand on a wet or damp floor. Never leave a running power tool unattended.

Tools that use compressed air are called pneumatic tools. Compressed air is used to inflate tires, apply paint, and drive tools. Compressed air can be dangerous when it is not used properly.

When using compressed air, safety glasses and/or a face shield should be worn. Particles of dirt and pieces of metal, blown by the high-pressure air, can penetrate your skin or get into your eyes.

Before using a compressed air tool, check all hose connections. Always hold an air nozzle or air control device securely when starting or shutting off the compressed air. A loose nozzle can whip suddenly and cause serious injury. Never point an air nozzle at anyone. Never use compressed air to blow dirt from your clothes or hair. Never use compressed air to clean the floor or workbench.

Always be careful when raising a vehicle on a lift or a hoist. Adapters and hoist plates must be positioned correctly to prevent damage to the underbody of the vehicle. There are specific lift points that allow the weight of the vehicle to be evenly supported by the adapters or hoist plates. The correct lift points can be found in the vehicle's service manual. Before operating any lift or hoist, carefully read the operating manual and follow the operating instructions.

Once you feel the lift supports are properly positioned under the vehicle, raise the lift until the supports contact the vehicle. Then, check the supports to make sure they are in full contact with the vehicle. Shake the vehicle to make sure it is securely balanced on the lift, and then raise the lift to the desired working height. Before working under a car, make sure the lift's locking devices are engaged.

A vehicle can be raised off the ground by a hydraulic jack. The jack's lifting pad must be positioned under an area of the vehicle's frame or at one of the manufacturer's recommended lift points. Never place the pad under the floor pan or under steering and suspension components, because these are easily damaged by the weight of the vehicle. Always position the jack so the wheels of the vehicle can roll as the vehicle is being raised.

Safety stands, also called jack stands, should be placed under a sturdy chassis member, such as the frame or axle housing, to support the vehicle after it has been raised by a jack. Once the safety stands are in position, the hydraulic pressure in the jack should be slowly released until the weight of the vehicle is on the stands. Never move under a vehicle when it is supported only by a hydraulic jack. Rest the vehicle on the safety stands before moving under the vehicle.

Heavy parts of the automobile, such as engines, are removed with chain hoists or cranes. Cranes often are called cherry pickers. To prevent serious injury, chain hoists and cranes must be properly attached to the parts being lifted. Always use bolts with enough strength to support the object being lifted. After you have attached the lifting chain or cable to the part that is being removed, have your instructor check it. Place the chain hoist or crane directly over the assembly, and then attach the chain or cable to the hoist.

Parts cleaning is a necessary step in most repair procedures. Always wear the appropriate protection when using chemical, abrasive, and thermal cleaners.

Vehicle Operation

When the customer brings a vehicle in for service, certain driving rules should be followed to ensure your safety and the safety of those working around you. For example, before moving a car into the shop, buckle your safety belt. Make sure no one is near, the way is clear, and there are no tools or parts under the car before you start the engine. Check the brakes before putting the vehicle in gear. Then, drive slowly and carefully in and around the shop.

If the engine must be running while you are working on the car, block the wheels to prevent the car from moving. Place the transmission into park for automatic transmissions or into neutral for manual transmissions. Set the parking (emergency) brake. Never stand directly in front of or behind a running vehicle.

Run the engine only in a well-ventilated area to avoid the danger of poisonous carbon monoxide (CO) in the engine exhaust. CO is an odorless but deadly gas. Most shops have an exhaust ventilation system, and you should always use it. Connect the hose from the vehicle's tailpipe to the intake for the vent system. Make sure the vent system is turned on before running the engine. If the work area does not have an exhaust venting system, use a hose to direct the exhaust out of the building.

HAZARDOUS MATERIALS AND WASTES

A typical shop contains many potential health hazards for those working in it. These hazards can cause injury, sickness, health impairments, discomfort, and even death. Here is a short list of the different classes of hazards:

- Chemical hazards are caused by high concentrations of vapors, gases, or solids in the form of dust.
- Hazardous wastes are those substances that result from a service being performed.
- Physical hazards include excessive noise, vibration, pressure, and temperature.
- Ergonomic hazards are conditions that impede normal and/or proper body position and motion.

There are many government agencies charged with ensuring safe work environments for all workers. These include the Occupational Safety and Health Administration (OSHA), Mine Safety and Health Administration (MSHA), and National Institute for Occupational Safety and Health (NIOSH). These, as well as state and local governments, have instituted regulations that must be understood and followed. Everyone in a shop has the responsibility for adhering to these regulations.

An important part of a safe work environment is the employees' knowledge of potential hazards. Right-to-know laws concerning all chemicals protect every employee in the shop. The general intent of right-to-know laws is to ensure that employers provide their employees with a safe working place as far as hazardous materials are concerned.

All employees must be trained about their rights under the legislation, the nature of the hazardous chemicals in their workplace, and the contents of the labels on the chemicals. All of the information about each chemical must be posted on material safety data sheets (MSDS) and must be accessible. The manufacturer of the chemical must give these sheets to its customers, if they are requested to do so. The sheets detail the chemical composition and precautionary information for all products that can present a health or safety hazard.

Employees must become familiar with the general uses, protective equipment, accident or spill procedures, and any other information regarding the safe handling of a particular hazardous material. This training must be given to employees annually and provided to new employees as part of their job orientation.

All hazardous material must be properly labeled, indicating what health, fire, or reactivity hazard it poses and what protective equipment is necessary when handling each chemical. The manufacturer of the hazardous materials must provide all warnings and precautionary information, which must be read and understood by the user before use. A list of all hazardous materials used in the shop must be posted for the employees to see.

Shops must maintain documentation on the hazardous chemicals in the workplace, proof of training programs, records of accidents or spill incidents, satisfaction of employee requests for specific chemical information via the MSDS, and a general right-to-know compliance procedure manual utilized within the shop.

When handling any hazardous materials or hazardous waste, make sure you follow the required procedures for handling such material. Also wear the proper safety equipment listed on the MSDS. This includes the use of approved respirator equipment.

Some of the common hazardous materials that automotive technicians use are: cleaning chemicals, fuels (gasoline and diesel), paints and thinners, battery electrolyte (acid), used engine oil, refrigerants, and engine coolant (anti-freeze).

Many repair and service procedures generate what are known as hazardous wastes. Dirty solvents and cleaners are good examples of hazardous wastes. Something is classified as a hazardous waste if it is on the EPA list of known harmful materials or has one or more of the following characteristics.

- *Ignitability.* If it is a liquid with a flash point below 140° F or a solid that can spontaneously ignite.
- *Corrosivity.* If it dissolves metals and other materials or burns the skin.
- *Reactivity.* Any material that reacts violently with water or other materials or releases cyanide gas, hydrogen sulfide gas, or similar gases when exposed to low pH acid solutions. This also includes material that generates toxic mists, fumes, vapors, and flammable gases.
- *Toxicity.* Materials that leach one or more of eight heavy metals in concentrations greater than 100 times primary drinking water standard concentrations.

Complete EPA lists of hazardous wastes can be found in the Code of Federal Regulations. It should be noted that no material is considered hazardous waste until the shop is finished using it and ready to dispose of it.

The following list covers the recommended procedure for dealing with some of the common hazardous wastes. Always follow these and any other mandated procedures.

Oil   Recycle oil. Set up equipment, such as a drip table or screen table with a used oil collection bucket, to collect oils dripping off parts. Place drip pans underneath vehicles that are leaking fluids onto the storage area. Do not mix other wastes with used oil, except as allowed by your recycler. Used oil generated by a shop (and/or oil received from household "do-it-yourself" generators) may be burned on site in a commercial space heater. Also, used oil may be burned for energy recovery. Contact state and local authorities to determine requirements and to obtain the necessary permits.

Oil filters   Drain for at least 24 hours, crush, and recycle used oil filters.

Batteries   Recycle batteries by sending them to a reclaimer or back to the distributor. Keeping shipping receipts can demonstrate that you have done the recycling. Store batteries in a watertight, acid-resistant container. Inspect batteries for cracks and leaks when they come in. Treat a dropped battery as if it were cracked. Acid residue is hazardous because it is corrosive and may contain lead and other toxic substances. Neutralize spilled acid, by using baking soda or lime, and dispose of it as hazardous material.

Metal residue from machining   Collect metal filings when machining metal parts. Keep them separate and recycle if possible. Prevent metal filings from falling into a storm sewer drain.

Refrigerants   Recover and/or recycle refrigerants during the servicing and disposal of motor vehicle air conditioners and refrigeration equipment. It is not allowable to knowingly vent refrigerants to the atmosphere. Recovering and/or recycling must be performed by an EPA-certified technician using certified equipment and following specified procedures.

Solvents   Replace hazardous chemicals with less toxic alternatives that perform equally. For example, substitute water-based cleaning solvents for petroleum-based solvent degreasers. To reduce the amount of solvent used when cleaning parts, use a two-stage process: dirty solvent followed by fresh solvent. Hire a hazardous waste management service to clean and recycle solvents. (Some spent solvents must be disposed of as hazardous waste, unless recycled properly). Store solvents in closed containers to prevent evaporation. Evaporation of solvents contributes to ozone depletion and smog formation. In addition, the residue from evaporation must be treated as a hazardous waste. Properly label spent solvents and store them on drip pans or in diked areas and only with compatible materials.

Containers   Cap, label, cover, and properly store aboveground outdoor liquid containers and small tanks within a diked area and on a paved impermeable surface to prevent spills from running into surface or ground water.

Other solids   Store materials such as scrap metal, old machine parts, and worn tires under a roof or tarpaulin to protect them from the elements and to prevent the possibility of creating contaminated runoff. Consider recycling tires by retreading them.

Liquid recycling   Collect and recycle coolants from radiators. Store transmission fluids, brake fluids, and solvents containing chlorinated hydrocarbons separately, and recycle or dispose of them properly.

Shop towels and rags   Keep waste towels in a closed container marked "contaminated shop towels only." To reduce costs and liabilities associated with disposal of used towels, which can be classified as hazardous wastes, investigate using a laundry service that is able to treat the wastewater generated from cleaning the towels.

Waste storage   Always keep hazardous waste separate, properly labeled, and sealed in the recommended containers. The storage area should be covered and may need to be fenced and locked if vandalism could be a problem. Select a licensed hazardous waste hauler after seeking recommendations and reviewing the firm's permits and authorizations.

NATEF TASK LIST FOR ENGINE REPAIR

A. General Engine Diagnosis; Removal and Reinstallation (R&R)

A.1. Identify and interpret engine concern; determine necessary action. Priority Rating 1

A.2. Research applicable vehicle and service information, such as internal engine operation, vehicle service history, service precautions, and technical service bulletins.

A.3. Locate and interpret vehicle and major component identification numbers (VIN, vehicle certification labels, and calibration decals). Priority Rating 1

A.4. Inspect engine assembly for fuel, oil, coolant, and other leaks; determine necessary action. Priority Rating 2

A.5. Diagnose engine noises and vibrations; determine necessary action. Priority Rating 3

A.6. Diagnose the cause of excessive oil consumption, unusual engine exhaust color, odor, and sound; determine necessary action. Priority Rating 3

A.7. Perform engine vacuum tests; determine necessary action. Priority Rating 1

A.8. Perform cylinder power balance tests; determine necessary action. Priority Rating 1

A.9. Perform cylinder compression tests; determine necessary action. Priority Rating 1

A.10. Perform cylinder leakage tests; determine necessary action. Priority Rating 1

A.11. Remove and reinstall engine in a late model front-wheel-drive vehicle (OBD-I or newer); reconnect all attaching components and restore the vehicle to running condition. Priority Rating 3

A.12. Remove and reinstall engine in a late model rear-wheel-drive vehicle (OBD-I or newer); reconnect all attaching components and restore the vehicle to running condition. Priority Rating 3

B. Cylinder Head and Valve Train Diagnosis and Repair

B.1. Remove cylinder head(s); visually inspect cylinder head(s) for cracks; check gasket surface areas for warpage and leakage; check passage condition. Priority Rating 2

B.2. Install cylinder heads and gaskets; tighten according to manufacturer's specifications and procedures. Priority Rating 2

B.3. Inspect and test valve springs for squareness, pressure, and free height comparison; determine necessary action. Priority Rating 3

B.4. Replace valve stem seals on an assembled engine; inspect valve retainers, locks, and valve grooves; determine necessary action. Priority Rating 2

B.5. Inspect valve guides for wear; check valve guide height and stem-to-guide clearance; determine necessary action. Priority Rating 3

B.6. Inspect valves and valve seats; determine necessary action.

B.7. Check valve face-to-seat contact and valve seat concentricity (runout); determine necessary action. Priority Rating 3

B.8. Check valve spring assembled height and valve stem height; determine necessary action. Priority Rating 2

B.9. Inspect pushrods, rocker arms, rocker arm pivots and shafts for wear, bending, cracks, looseness, and blocked oil passages (orifices); determine necessary action. Priority Rating 2

B.10. Inspect hydraulic or mechanical lifters; determine necessary action. Priority Rating 2

B.11. Adjust valves (mechanical or hydraulic lifters). Priority Rating 1

B.12. Inspect camshaft drives (including gear wear and backlash, sprocket and chain wear); determine necessary action. Priority Rating 2

B.13. Inspect and replace timing belts(s), overhead camdrive sprockets, and tensioners; check belt/chain tension; adjust as necessary. Priority Rating 1

B.14. Inspect camshaft for runout, journal wear and lobe wear. Priority Rating 3

B.15. Inspect camshaft bearing surface for wear, damage, out-of-round, and alignment; determine necessary action. Priority Rating 3

B.16. Establish camshaft(s) timing and cam sensor indexing according
to manufacturer's specifications and procedure.   Priority Rating 1

C. Engine Block Assembly Diagnosis and Repair

C.1. Disassemble engine block; clean and prepare components for
inspection and reassembly.

C.2. Inspect engine block for visible cracks, passage condition, core and
gallery plug condition, and surface warpage; determine necessary action.   Priority Rating 2

C.3. Inspect internal and external threads; restore as needed
(includes installing thread inserts).   Priority Rating 1

C.4. Inspect and measure cylinder walls for damage and wear; determine
necessary action.   Priority Rating 2

C.5. Deglaze and clean cylinder walls.   Priority Rating 1

C.6. Inspect and measure camshaft bearings for wear, damage,
out-of-round, and alignment; determine necessary action.   Priority Rating 3

C.7. Inspect crankshaft for end play, straightness, journal damage,
keyway damage, thrust flange and sealing surface condition, and
visual surface cracks; check oil passage condition; measure journal
wear; check crankshaft sensor reluctor ring (where applicable).   Priority Rating 3

C.8. Inspect and measure main and connecting rod bearings for
damage, clearance, and end play; determine necessary action
(includes the proper selection of bearings).   Priority Rating 2

C.9. Identify piston and bearing wear patterns that indicate connecting
rod alignment and main bearing bore problems; inspect rod
alignment and bearing bore condition.   Priority Rating 3

C.10. Inspect and measure pistons; determine necessary action.   Priority Rating 2

C.11. Remove and replace piston pins.   Priority Rating 2

C.12. Inspect, measure, and install piston rings.   Priority Rating 2

C.13. Inspect auxiliary (balance, intermediate, idler, counterbalance or
silencer) shaft(s); inspect shaft(s) and support bearings for
damage and wear; determine necessary action; reinstall and time. Priority Rating 3

C.14. Inspect, repair or replace crankshaft vibration damper
(harmonic balancer).   Priority Rating 3

C.15. Assemble the engine using gaskets, seals, and formed-in-place
(tube-applied) sealants, thread sealers, etc. according to
manufacturer's specifications.   Priority Rating 2

D. Lubrication and Cooling Systems Diagnosis and Repair

D.1. Perform oil pressure tests; determine necessary action.   Priority Rating 1

D.2. Inspect oil pump gears or rotors, housing, pressure relief
devices, and pump drive; perform necessary action.   Priority Rating 3

D.3. Perform cooling system, cap, and recovery system tests (pressure,
combustion leakage, and temperature); determine necessary action. Priority Rating 1

D.4. Inspect, replace, and adjust drive belts, tensioners, and pulleys;
check pulley and belt alignment.   Priority Rating 1

D.5. Inspect and replace engine cooling and heater system hoses. Priority Rating 2

D.6. Inspect, test, and replace thermostat and housing.   Priority Rating 2

D.7. Test coolant; drain and recover coolant; flush and refill cooling
system with recommended coolant; bleed air as required.   Priority Rating 1

D.8. Inspect, test, remove, and replace water pump.   Priority Rating 2

D.9. Remove and replace radiator.   Priority Rating 2

D.10. Inspect, and test fans(s) (electrical or mechanical), fan clutch,
fan shroud, and air dams.   Priority Rating 2

D.11. Inspect auxiliary oil coolers; determine necessary action. Priority Rating 3

D.12. Inspect, test, and replace oil temperature and pressure switches
and sensors.   Priority Rating 2

D.13. Perform oil and filter change.   Priority Rating 1

DEFINITION OF TERMS USED IN THE TASK LIST

To clarify the intent of these tasks, NATEF has defined some of the terms used in the task listings. To get a good understanding of what the task includes, refer to this glossary while reading the task list.

add	To increase fluid or pressure to the correct level or amount.
adjust	To bring components to specified operational settings.
assemble (reassemble)	To fit together the components of a device.
check	To verify condition by performing an operational or comparative examination.
clean	To rid components of extraneous matter for the purpose of reconditioning, repairing, measuring and reassembling.
deglaze	To restore correct surface finish.
determine	To establish the procedure to be used to effect the necessary repair.
determine necessary action	Indicates that the diagnostic routine(s) is the primary emphasis of a task. The student is required to perform the diagnostic steps and communicate the diagnostic outcomes and corrective actions required addressing the concern or problem. The training program determines the communication method (worksheet, test, verbal communication, or other means deemed appropriate) and whether the corrective procedures for these tasks are actually performed.
diagnose	To locate the root cause or nature of a problem by using the specified procedure.
disassemble	To separate a component's parts as a preparation for cleaning, inspection or service.
drain	To use gravity to empty a container.
fill (refill)	To bring fluid level to specified point or volume.
flush	To use fluid to clean an internal system.
identify	To establish the identity of a vehicle or component prior to service; to determine the nature or degree of a problem.
inspect	(see *check*)
install (reinstall)	To place a component in its proper position in a system.
measure	To compare existing dimensions to specified dimensions by the use of calibrated instruments and gauges.
perform	To accomplish a procedure in accordance with established methods and standards.
perform necessary action	Indicates that the student is to perform the diagnostic routine(s) and perform the corrective action item. Where various scenarios (conditions or situations) are presented in a single task, at least one of the scenarios must be accomplished.
pressure test	To use air or fluid pressure to determine the integrity, condition, or operation of a component or system.
reassemble	(see *assemble*)
refill	(see *fill*)
remove	To disconnect and separate a component from a system.
repair	To restore a malfunctioning component or system to operating condition.
replace	To exchange an unserviceable component with a new or rebuilt component; to reinstall a component.

select	To choose the correct part or setting during assembly or adjustment.
service	To perform a specified procedure when called for in the owner's or service manual.
test	To verify condition through the use of meters, gauges, or instruments.
torque	To tighten a fastener to specified degree of tightness (in a given order or pattern if multiple fasteners are involved on a single component).
vacuum test	To determine the integrity and operation of a vacuum (negative pressure) operated component and/or system.
verify	To establish that a problem exists after hearing the customer's complaint and performing a preliminary diagnosis.

ENGINE REPAIR TOOLS AND EQUIPMENT

Many different tools and many kinds of testing and measuring equipment are used to service engines. NATEF has identified many of these and has said an Engine Repair technician must know what they are and how and when to use them. The tools and equipment listed by NATEF are covered in the following discussion. Also included are the tools and equipment you will use while completing the job sheets. Although you need to be more than familiar with and will be using common hand tools, they are not part of this discussion. You should already know what they are and how to use and care for them.

Compression Testers

Engines depend on the compression of the air/fuel mixture to have power output. The compression stroke of the piston compresses the air/fuel mixture within the combustion chamber. If the combustion chamber leaks, some of the air/fuel mixture will escape while it is being compressed, resulting in a loss of power and a waste of fuel. The leaks can be caused by burned valves, a blown head gasket, worn rings, slipped timing belt or chain, worn valve seats, a cracked head, and other factors.

An engine with poor compression (lower compression pressure due to leaks in the cylinder) will not run correctly. If a symptom suggests that the cause of a problem may be poor compression, a compression test is performed.

A compression gauge is used to check cylinder compression. The dial face on the typical compression gauge indicates pressure in both pounds per square inch (psi) and metric kilopascals (kPa). The range is usually 0 to 300 psi and 0 to 2,100 kPa. There are two basic types of compression gauges: a push-in gauge and a screw-in gauge.

The push-in type has a short stem that is either straight or bent at a 45-degree angle. The stem ends in a tapered rubber tip that fits any size spark plug hole. The rubber tip is placed in the spark plug hole, after the spark plugs have been removed, and held there while the engine is cranked through several compression cycles. Although simple to use, the push-in gauge may give inaccurate readings if it is not held tightly in the hole.

The screw-in gauge has a long, flexible hose that ends in a threaded adapter. This type compression tester is often used because its flexible hose can reach into areas that are difficult to reach with a push-in type tester. The threaded adapters are changeable and come in several thread sizes to fit 10-mm, 12-mm, 14-mm, and 18-mm diameter holes. The adapters screw into the spark plug holes in place of the spark plugs.

Most compression gauges have a vent valve that allows the gauge's meter to hold the highest pressure reading. Opening the valve releases the pressure when the test is complete.

Cylinder Leakage Tester

If a compression test shows that any of the cylinders are leaking, a cylinder leakage test can be performed to measure the percentage of compression lost and help locate the source of leakage.

A cylinder leakage tester applies compressed air to a cylinder through the spark plug hole. Before the air is applied to the cylinder, the piston of that cylinder must be at TDC on its compression stroke. A threaded adapter on the end of the air pressure hose screws into the spark plug hole. The source of the compressed air is normally the shop's compressed air system. A pressure regulator in the tester controls the pressure applied to the cylinder. An analog gauge registers the percentage of air pressure lost from the cylinder when the compressed air is applied. The scale on the dial face reads 0 to 100 percent.

A zero reading means there is no leakage from the cylinder. Readings of 100 percent would indicate that the cylinder does not hold pressure. The location of the compression leak can be found by listening and feeling around various parts of the engine. If air is felt or heard leaving the throttle plate assembly, a leaking intake valve is indicated. If a bad exhaust valve is responsible for the leakage, air can be felt leaving the exhaust system during the test. Air leaving the radiator would indicate a faulty head gasket or a cracked block or head. If the piston rings are bad, air will be heard leaving the valve cover's breather cap or the oil dipstick tube.

Most vehicles, even new cars, experience some leakage around the rings. Up to 20 percent is considered acceptable during the leakage test. When the engine is actually running, the rings will seal much more tightly and the actual percent of leakage will be lower. However, there should be no leakage around the valves or the head gasket.

Vacuum Gauge

Measuring intake manifold vacuum is another way to diagnose the condition of an engine. Manifold vacuum is tested with a vacuum gauge. Vacuum is formed on a piston's intake stroke. As the piston moves down, it lowers the pressure of the air in the cylinder—if the cylinder is sealed. This lower cylinder pressure is called engine vacuum. If there is a leak, atmospheric pressure will force air into the cylinder and the resultant pressure will not be as low. The reason atmospheric pressure enters is simply that whenever there is a low and high pressure, the high pressure will always move toward the low pressure.

Vacuum is measured in inches of mercury (in./Hg) and in kilopascals (kPa) or millimeters of mercury (mm/Hg).

To measure vacuum, a flexible hose on the vacuum gauge is connected to a source of manifold vacuum, either on the manifold or at a point below the throttle plates. Sometimes this requires removing a plug from the manifold and installing a special fitting.

The test is made with the engine cranking or running. A good vacuum reading is typically at least 16 in./Hg; however, a reading of 15 to 20 in./Hg (50 to 65 kPa) is normally acceptable. Since the intake stroke of each cylinder occurs at a different time, the production of vacuum occurs in pulses. If the amount of vacuum produced by each cylinder is the same, the vacuum gauge will show a steady reading. If one or more cylinders are producing different amounts of vacuum, the gauge will show a fluctuating reading.

Oil Pressure Gauge

Checking the engine's oil pressure will give you information about the condition of the oil pump, pressure regulator, and the entire lubrication system. Lower than normal oil pressures can be caused by excessive engine bearing clearances. Oil pressure is checked at the sending unit passage with an externally mounted mechanical oil pressure gauge (Figure 2). Various fittings are usually supplied with the oil pressure gauge to fit different openings in the lubrication system.

To get accurate results from the test, make sure you follow the manufacturer's recommendations and compare your findings to specifications. Low oil pressure readings can be caused by internal component wear, pump-related problems, low oil level, contaminated oil, or low oil viscosity. An overfilled crankcase, high oil viscosity, or a faulty pressure regulator can cause high oil pressure readings.

Stethoscope

Some engine sounds can be easily heard without using a listening device, but others are impossible to hear unless they are amplified. A stethoscope is very helpful in locating engine noise because it amplifies the sound waves. It can also distinguish between normal and abnormal noise. The procedure for using a

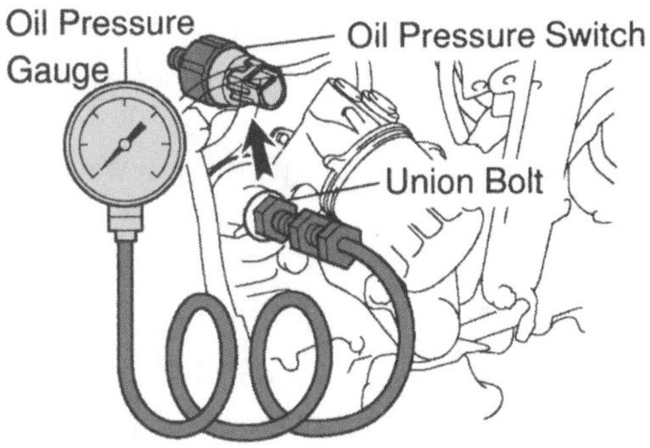

Oil Pressure
Gauge
Oil Pressure Switch
Union Bolt

Figure 2 An oil pressure gauge can be connected to the bore for the oil pressure switch or sending unit.

stethoscope is simple. Use the metal prod to trace the sound until it reaches its maximum intensity. Once the precise location has been discovered, the sound can be better evaluated. A sounding stick, which is nothing more than a long, hollow tube, works on the same principle, though a stethoscope gives much clearer results.

The best results, however, are obtained with an electronic listening device. With this tool you can tune into the noise. Doing this allows you to eliminate all other noises that might distract or mislead you.

Portable Crane

To remove and install an engine, a portable crane, frequently called a cherry picker, is used. To lift an engine, attach a pulling sling or chain to the engine. Some engines have eye plates for use in lifting. If they are not available, the sling must be bolted to the engine. The sling attaching bolts must be large enough to support the engine and must thread into the block a minimum of 1-1/2 times the bolt diameter. Connect the crane to the chain. Raise the engine slightly and make sure the sling attachments are secure. Carefully lift the engine out of its compartment.

Lower the engine close to the floor so the transmission and torque converter or clutch can be removed from the engine, if necessary.

Engine Stands/Benches

After the engine has been removed, use the crane to raise the engine. Position the engine next to an engine stand. Most stands use a plate with several holes or adjustable arms. The engine must be supported by at lease four bolts that fit solidly into the engine. The engine should be positioned so that its center is in the middle of the engine's stand adapter plate (Figure 3). The adapter plate can swivel in the stand. By centering the engine, it can be easily turned to the desired working positions.

Figure 3 A cylinder block mounted to an engine stand.

Some shops have engine mounts bolted to the top of workbenches. The engine is suspended off the side of the workbench. This kind of workbench has the advantage of a good working space next to the engine, but it is not mobile and all engine work must be done at that location.

After the engine is secured to its mount, the crane and lifting chains can be removed and disassembly of the engine can begin.

Transaxle Removal and Installation Equipment

R&R of transversely mounted engines may require other tools. The engines of some FWD vehicles are removed by lifting them from the top. Others must be removed from the bottom and the procedure requires different equipment. Make sure you follow the instructions given by the manufacturer and use the appropriate tools and equipment. The required equipment varies with manufacturer and vehicle model; however, most accomplish the same thing.

To remove the engine from under the vehicle, the vehicle must be raised. A crane and/or support fixture is used to hold the engine and transaxle assembly in place while the engine is being readied for removal. Once the engine is ready, the crane is used to lower the engine onto an engine cradle. The cradle is similar to a hydraulic floor jack and is used to lower the engine to the point where it can be rolled out from under the vehicle.

Often a transverse-mounted engine is removed with the transaxle, as a unit. The transaxle can be separated from the engine once it has been lifted out of the vehicle. In this case, the drive axles must be disconnected from the transaxle before removing the unit.

Machinist's Rule

A machinist's rule is very much like an ordinary ruler. Each edge of this measuring tool is divided into increments based on a different scale. A typical machinist's rule based on the USCS system of measurement may have scales based on 1/8-, 1/16-, 1/32-, and 1/64-inch intervals. Of course, metric machinist rules are also available. Metric rules are usually divided into 0.5-mm and 1-mm increments.

Some machinist rules are based on decimal intervals. These are typically divided into 1/10-, 1/50-, and 1/1,000-inch (0.1, 0.01, and 0.001) increments. Decimal machinist rules are very helpful when measuring dimensions that are specified in decimals; they eliminate the need to convert fractions to decimals.

Micrometers

A micrometer is used to measure linear outside and inside dimensions. Both outside and inside micrometers are calibrated and read in the same manner. The major components and markings of a micrometer include the frame, anvil, spindle, locknut, sleeve, sleeve numbers, sleeve long line, thimble marks, thimble, and ratchet. Micrometers are calibrated in either inch or metric graduations and are available in a range of sizes.

To use and read a micrometer, choose the appropriate size for the object being measured. Typically they measure an inch, therefore the range covered by a micrometer of one size would be from 0 to 1 inch and another would measure 1 to 2 inches, and so on.

Open the jaws of the micrometer and slip the object between the spindle and anvil (Figure 4). While holding the object against the anvil, turn the thimble using your thumb and forefinger until the spindle contacts the object. Never clamp the micrometer tightly. Use only enough pressure on the thimble to allow the work to just fit between the anvil and spindle. To get accurate readings, you should slip the micrometer back and forth over the object until you feel a very light resistance, while at the same time rocking the tool from side to side to make certain the spindle cannot be closed any further. When a satisfactory adjustment has been made, lock the micrometer and read the measurement scale.

The graduations on the sleeve each represent 0.025 inch. To read a measurement on a micrometer, begin by counting the visible lines on the sleeve and multiplying them by 0.025. The graduations on the thimble assembly define the area between the lines on the sleeve. The number indicated on the thimble is added to the measurement shown on the sleeve. This sum is the dimension of the object.

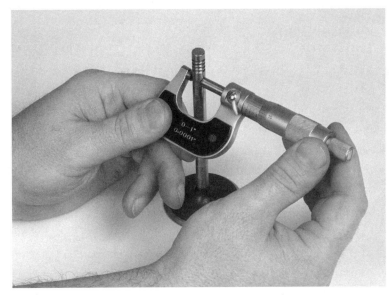

Figure 4 Using a micrometer to measure a valve stem.

Micrometers are available that measure in 0.0001 (ten-thousandths) of an inch. Use this type of micrometer if the specifications call for this degree of accuracy.

A metric micrometer is read in the same way, except the graduations are expressed in the metric system of measurement. Each number on the sleeve represents 5 millimeters (mm) or 0.005 meter (m). Each of the ten equal spaces between each number, with index lines alternating above and below the horizontal line, represents 0.5 mm or five tenths of an mm. Therefore, one revolution of the thimble changes the reading one space on the sleeve scale or 0.5 mm. The beveled edge of the thimble is divided into 50 equal divisions with every fifth line numbered consecutively 0, 5, 10, . . ., 45. Since one complete revolution of the thimble advances the spindle 0.5-mm, each graduation on the thimble is equal to one hundredth of a millimeter. As with the micrometer that is graduated in inches, the separate readings are added together to obtain the total reading.

Some technicians use a digital micrometer, which is easier to read. These tools do not have the various scales; instead, the measurement is displayed and read directly off the micrometer.

Inside micrometers can be used to measure the inside diameter of a bore. To do this, place the tool inside the bore and extend the measuring surfaces until each end touches the bore's surface. If the bore is large, it might be necessary to use an extension rod to increase the micrometer's range. These extension rods come in various lengths. The inside micrometer is read in the same manner as an outside micrometer.

A depth micrometer is used to measure the distance between two parallel surfaces. The sleeves, thimbles, and ratchet screws operate in the same way as other micrometers. Likewise, depth micrometers are read in the same way as other micrometers. If a depth micrometer is used with a gauge bar, it is important to keep both the bar and the micrometer from rocking. Any movement of either part will result in an inaccurate measurement.

Telescoping Gauge

Telescoping gauges are used for measuring bore diameters and other clearances (Figure 5). They may also be called snap gauges. They are available in sizes ranging from fractions of an inch through 6 inches. Each gauge consists of two telescoping plungers, a handle, and a lock screw. Snap gauges are normally used with an outside micrometer.

To use the telescoping gauge, insert it into the bore and loosen the lock screw. This will allow the plungers to snap against the bore. Once the plungers have expanded, tighten the lock screw. Then, remove the gauge and measure the expanse with a micrometer.

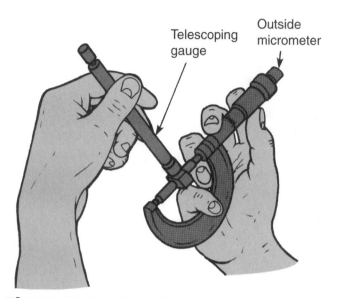

Figure 5 Measuring a telescoping gauge with an outside micrometer.

Small Hole Gauge

A small hole or ball gauge works just like a telescoping gauge. However, it is designed for small bores. After it is placed into the bore and expanded, it is removed and measured with a micrometer. Like the telescoping gauge, the small hole gauge consists of a lock, handle, and an expanding end. The end expands or retracts by turning the gauge handle.

Feeler Gauge

A feeler gauge is a thin strip of metal or plastic of known and closely controlled thickness. Several of these strips are often assembled together as a feeler gauge set that looks like a pocketknife. The desired thickness gauge can be pivoted away from the others for convenient use. A feeler gauge set usually contains strips or leaves of 0.002- to 0.010-inch thickness (in steps of 0.001 inch) and leaves of 0.012- to 0.024-inch thickness (in steps of 0.002 inch).

A feeler gauge can be used by itself to measure piston ring side clearance, piston ring end gap, connecting rod side clearance, crankshaft endplay, and other distances. It can also be used with a precision straightedge to measure main bearing bore alignment and cylinder head/block warpage.

Straightedge

A straightedge is no more than a flat bar machined to be totally flat and straight, and to be effective it must be flat and straight. Any surface that should be flat can be checked with a straightedge and feeler gauge set. The straightedge is placed across and at angles on the surface. At any low points on the surface, a feeler gauge can be placed between the straight edge and the surface. The size of the gauge that fills in the gap indicates the amount of warpage or distortion (Figure 6).

Dial Indicator

The dial indicator is calibrated in 0.001-inch (one-thousandth inch) increments. Metric dial indicators are also available. Both types are used to measure movement. Common uses of the dial indicator include measuring valve lift, journal concentricity, flywheel or brake rotor runout, gear backlash, and crankshaft endplay.

To use a dial indicator, position the indicator rod against the object to be measured. Then, push the indicator toward the work until the indicator needle travels far enough around the gauge face to permit movement to be read in either direction. Zero the indicator needle on the gauge. Move the object in the direction required, while observing the needle of the gauge. Always be sure the range of the dial indica-

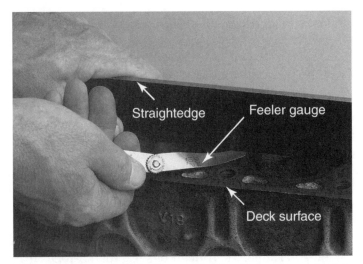

Figure 6 Using a feeler gauge and straightedge to check for warpage.

tor is sufficient to allow the amount of movement required by the measuring procedure. For example, never use a 1-inch indicator on a component that will move 2 inches.

Dial Bore Gauge

Cylinder bore taper and out-of-roundness can be measured with a micrometer and telescoping gauge. However, most shops use a dial bore gauge, which typically consists of a handle, guide blocks, lock, indicator contact, and an indicator. It also comes with extensions that make it adaptable to various sized bores.

Most cylinder wear occurs at the top of the ring travel area. Pressure on the top ring is at a peak and lubrication at a minimum when the piston is at the top of its stroke. A ridge of unworn material will remain above the upper limit of ring travel. Below the ring travel area, wear is negligible because only the piston skirt contacts the cylinder wall.

A properly reconditioned cylinder must have the correct diameter, have no taper or out-of-roundness, and the surface finish must be such that the piston rings will seat to form a seal that will control oil and minimize blowby.

Taper is the difference in diameter between the bottom of the cylinder bore and the top of the bore just below the ridge. Subtracting the smaller diameter from the larger one gives the cylinder taper. On a dial bore gauge, the change in diameter will move the indicator. The highest reading on the indicator is the amount of taper in the bore.

Cylinder out-of-roundness is the difference between the cylinder's diameter when measured parallel with the crank and then perpendicular to the crank. Out-of-roundness is measured at the top of the cylinder just below the ridge. As the dial bore gauge is turned inside the bore, the indicator will show any change in the bore's diameter. The amount shown on the indicator is the amount the cylinder is out-of-round.

Ridge Reamer

After many miles of use, a ridge is formed at the top of the engine's cylinders. Because the top piston ring stops traveling before it reaches the top of the cylinder, a ridge of unworn metal is left. This ridge must be removed to remove the pistons from the block without damaging them.

This ridge is removed with a ridge reamer or ridge-removing tool. The tool must be adjusted for the bore and then inserted into it. Rotate the tool clockwise with a wrench to remove the ridge. Remove just enough metal to allow the piston assembly to slip out of the bore without causing damage to the surface of the bore or to the piston. If the ridge is too large, the top rings will hit it and possibly break the ring lands.

After the ridge removing operation, wipe all the metal cuttings out of the cylinder. Use an oily rag to wipe the cylinder. The cuttings will stick to it.

Cylinder Deglazer

The proper surface finish on a cylinder wall acts as a reservoir for oil to lubricate the piston rings and prevent piston and ring scuffing. Piston ring faces can be damaged and experience premature wear if the cylinder wall is too rough. A surface that is too smooth will not hold enough oil and won't allow the rings to seat properly. Obtaining the correct cylinder wall finish is important.

If the inspection and measurements of the cylinder wall show that surface conditions, taper, and out-of-roundness are within acceptable limits, the cylinder walls only need to be deglazed. Combustion heat, engine oil, and piston movement combine to form a thin residue on the cylinder walls that is commonly called glaze.

The common types of glaze breakers use an abrasive with about 220 or 280 grit. The glaze breaker is installed in a slow-moving electric drill or in a honing machine. Many deglazers use round stones that extend on coiled wire from the center shaft (Figure 7). This type deglazer may also be used to lightly hone the bore. Various sizes of resilient-based hone-type brushes are available for honing and deglazing.

Cylinder Hone

You should hone a cylinder whenever there are minor problems with the bore. Honing will sand the walls to remove imperfections. A cylinder hone usually consists of two or three stones. The hone rotates at a selected speed and is moved up and down the cylinder's bore. The stones have outward pressure on them and remove some metal from the bore as they rotate within it. Honing oil flows over the stones and onto the cylinder wall to control the temperature and flush out any metallic and abrasive residue. The correct stones should be used to ensure that the finished walls have the correct surface finish. Honing stones are classified by grit size; typically the lower the grit number, the coarser the stone.

Figure 7 A resilient-based, hone-type brush, commonly called a ball hone.

Cylinder honing machines are available in manual and automatic models. The major advantage of the automatic type is that it allows the technician to dial in the exact crosshatch angle needed.

When cylinder surfaces are badly worn or excessively scored or tapered, a boring bar is used to cut the cylinders for oversize pistons or sleeves. A boring bar leaves a pattern on the cylinder wall similar to uneven screw threads; therefore, you should hone the bore to the correct finish after it has been bored.

Ring Groove Cleaner

Before installing piston rings onto a piston, the ring grooves should be cleaned. The carbon and other debris that may be present in the back of the groove will not allow the rings to compress evenly and completely into the grooves. Piston ring grooves are best cleaned with a ring groove cleaner, which is adjustable to fit the width and depth of the groove. Make sure it is properly adjusted before using it and make sure you do not damage the piston while cleaning it.

Ring Expander

To prevent damage to the piston rings during removal and installation, a ring expander should be used. To install a piston ring, the ring must be made large enough to fit over the piston. When this is done by hand, there is a chance the ring may become distorted because it is almost impossible to evenly expand the ring with your hands. The ring expander also reduces the chances of scratching or damaging the piston while working the ring into its proper location. The rings fit into the jaws of the expander and the handle of the tool is squeezed to expand the ring. Expand the rings only to the point where they can fit over the piston.

Ring Compressor

The use of a ring compressor is the only sensible way to install a piston with piston rings into a cylinder. The compressor wraps around the rings to make their outside diameter smaller than the inside diameter of the bore (Figure 8). With the compressor tool adjusted properly, the piston assembly can be easily pushed into the bore without damaging the bore or piston.

Figure 8 A piston ring compressor.

V-Blocks

The various shafts in an engine must be straight and free of distortion. Visually it is impossible to see any distortions unless the shaft is severely damaged. Warped or distorted shafts will cause many problems, including premature wear of the bearings they ride on. The best way to check a shaft is to place the ends of the shaft onto V-blocks. These blocks will support the shaft and allow you to rotate the shaft. Place the plunger of a dial indicator on the journals of the shaft and rotate the shaft. Any movement of the indicator's needle suggests a problem.

Cam Bearing Driver Set

The camshaft is supported by several friction-type bearings, or bushings. They are designed as one piece and are typically pressed into the camshaft bore in the cylinder head or block; however, some OHC engines use split bearings to support the camshaft. Camshaft bearings are normally replaced during engine rebuilding. The old bearings should be inspected for signs of unusual wear that may indicate an oiling or bore alignment problem.

Cam bearings are normally press fit into the block or head using a bushing driver and hammer. After the cam bearings have been installed, the oil holes in the bearings should be checked for proper alignment with those in the block or head. This will ensure correct oil supply to vital engine areas. Proper alignment can be checked by inserting a wire through the holes or by squirting oil into the holes. If the oil does not run out, the holes are misaligned. This procedure should be repeated with each bearing.

Valve Spring Compressor

To remove the valves from a cylinder head, the valve spring assemblies must be removed first. To do this, the valve spring must be compressed enough to remove the valve keepers, then the retainer. There are many types of spring compressors available. Some are designed to allow valve spring removal while the cylinder head is still on the engine block. Other designs are used only when the cylinder head is removed. There are also spring compressor tools designed for specific OHC and DOHC engines.

The pry bar-type compressor is used when installing valve oil seals when the cylinder head is still mounted to the block. With the cylinder's piston at TDC, shop air is fed into the cylinder to hold the valve up and prevent it falling into the cylinder. The pry bar is then used to compress the valve spring so the valve keepers can be removed.

Some OHC engines require the use of a special spring compressor. Often these special tools can be used when the cylinder head is attached to the block and when it is on a bench. These compressors bolt to the cylinder head and have a threaded plunger that fits onto the retainer. As the plunger is tightened down on the retainer, the spring compresses.

C-clamp-type valve compressors can only be used on cylinder heads after they have been removed from the block. This kind of compressor usually is a universal tool with interchangeable jaws. The spring is compressed either pneumatically or manually after the compressor is in place. One end of the clamp is positioned on the valve head and the other on the valve's retainer. After the compressor is adjusted, the compressor is activated to squeeze down on the spring. Once the spring is compressed, the valve keepers can be removed. Then the tension of the compressor is slowly released and the valve retainer and spring can be removed.

To reinstall the valve spring, insert the valve into the guide and position the valve spring inserts, valve spring, and retainer over the valve stem. With the spring compressor, compress the spring just enough to install the valve keepers into their grooves. Excess pressure may cause the retainer to damage the oil seal. Release the spring compressor and tap the valve stem with a rubber mallet to seat the keepers.

Valve Spring Tester

Before valve springs are used, they should be checked to make sure they are within specifications. Line up all of the springs on a flat surface and place a straightedge across the tops to check their freestanding height. Free height should be within 1/16 inch of manufacturer's specifications. Throw away any spring that is not within specifications.

A spring that is not square will cause side pressure on the valve stem and abnormal wear. To check squareness, set a spring upright against a carpenter's square. Rotate the spring until a gap appears between the spring and the square. Measure the gap with a feeler gauge. If the gap is more than 0.060 inch, the spring should be replaced.

Use a valve spring tester to check each valve's open and close pressure. Close pressure guarantees a tight seal. The open pressure overcomes valve train inertia and closes the valve when it should close. Spring tension must be checked at the installed spring height; therefore, if a shim is to be used, insert it under the spring on the valve spring tension gauge. Compress the spring into the installed height by pressing down on the tester's lever. The tester's gauge will reflect the pressure of the spring when compressed to the installed or valve closed height. Compare this reading to the specifications. Now compress the spring to the open height specification (Figure 9). Use the rule on the tester or a machinist rule to measure the compressed height. Read the pressure on the tester and compare this reading to the specifications. Any pressure outside the pressure range given in the specifications indicates the spring should be replaced. Low spring pressure may allow the valve to float during high-speed operation. Excessive spring tension may cause premature valve train wear. If the tension is within specifications, the spring can be installed on the valve stem.

Valve Guide Repair Tools

The amount of valve guide wear can be measured with a ball gauge and micrometer. Insert and expand the ball gauge at the top of the guide. Lock it to that diameter, remove it from the guide, and measure the ball gauge with an outside micrometer. Repeat this process with the ball gauge in the middle and the bottom of the guide. Compare your measurements to the specifications for valve guide inside diameter. Then compare your measurements against each other. Any difference in reading shows a taper or wear inside the guide.

Another way to check for excessive guide wear is with a dial indicator. The accuracy of this check is directly dependent on the amount the valve is open during the check. Some manufacturers specify this amount or provide special spacers that are installed over the valve stem to ensure the proper height. Attach the dial indicator to the cylinder head and position it so the plunger is at a right angle to the valve stem being measured. With the plunger in contact with the valve head, move the valve toward the indicator and set the dial indicator to zero. Now move the valve from the indicator. Observe the

Figure 9 A valve spring tester.

reading on the dial while doing this. The reading on the indicator is the total movement of the valve and is indicative of the guide's wear. Compare the reading to the specifications.

Valve guide wear or taper must be corrected. A certain minimum amount is needed for lubrication and thermal expansion of the valve stem. Exhaust valves require more clearance than intakes because they run hotter. Clearance should also be close enough to prevent a buildup of varnish and carbon deposits on the stems, which could cause sticking. Insufficient clearance, however, can lead to rapid stem and guide wear, scuffing, and sticking, which prevents the valve from seating fully. If the clearance is too great, oil control will be a serious problem.

Knurling is one of the fastest ways to restore the inside diameter dimensions of a worn valve guide. The procedure raises the surface of the guide ID by plowing tiny furrows through the surface of the metal. As the knurling tool cuts into the guide, metal is raised or pushed up on either side of the indentation. This effectively decreases the ID of the guide hole. A burnisher is used to make the ridges flat and produce the proper-sized hole to restore the correct guide-to-stem clearance.

Reaming is used to repair worn guides by increasing the guide hold size to take an oversize valve stem or by restoring the guide to its original diameter after installing inserts. When reaming a guide, limit the amount of metal removed and always reface the valve seat after the valve guide has been reamed.

Installing thin-wall guide liners is popular with many production engine rebuilders, as well as smaller shops. The liners are installed by first boring out the original guides to a specified amount and then pressing the liners into the guide. Some of these liners are not precut to length and the excess must be milled off before finishing.

Replacing the entire valve guide is another repair option that is possible on cylinder heads with replaceable guides. To replace integral guides, bore or drive the old guide out and drive a thin-wall replacement guide into the hole. It is important to keep the centerline of the guide concentric with the valve seat so that the rocker arm-to-valve stem contact area is not disturbed.

If the original guide can be removed and a new one inserted, press out an old valve guide. Do this by placing the properly sized driver into the guide. The shoulder on the driver must also be slightly smaller than the OD of the guide, so that it will go through the cylinder head. Then press out the guide. To install a new guide, use a press and the same driver that was used to remove the old guide. Align the new guide and press straight down, not at an angle.

Valve and Valve Seat Resurfacing Equipment

Whenever the valves have been removed from the cylinder head, the valve heads and valve seats should be resurfaced. The most critical sealing surface in the valve train is between the face of the valve and its seat in the cylinder head when the valve is closed. Leakage between these surfaces reduces the engine's compression and power, and can lead to valve burning. To insure proper seating of the valve, the seat area on the valve face and seat must be the correct width, at the correct location, and concentric with the guide. These conditions are accomplished by renewing the surface of the valve face and seat.

Valve grinding or refacing is done by machining a fresh, smooth surface on the valve faces and stem tips. Valve faces suffer from burning, pitting, and wear caused by opening and closing millions of times during the life of an engine. Valve stem tips wear because of friction from the rocker arms or actuators.

Before grinding, each valve face should be inspected for burning or distortion and each stem tip for wear. Replace any valves that are badly burned or worn. Valves can be refaced using either grinding or cutting machines. Although both can reface valves and smoothen and chamfer valve stem ends, the traditional grinding method of refacing is still the most popular.

To start the valve grinding operation, chuck the valve as close as possible to the valve head to eliminate stem flexing from wheel pressure. Set the grinding angle according to specifications or the desired angle. If an interference angle is specified, the valve grinder should be set at 1/2 to 1 degree less than the standard 30- or 45-degree face angle. Make light cuts using the full width of the grinding wheel width. Remove only enough metal to clean up the valve face.

Once the face is ground, the valve tip may also need to be ground. This is best determined by placing the valve in the cylinder head to check the stem height. The tip is ground by using a stemming stone with the valve secured and a coolant flowing over the stem to cool the valve tip and remove grit. The valve tip should be ground so that it is exactly square with the stem. Because valve tips have hardened surfaces that are up to 0.030 inch in depth, only 0.010 inch can be removed during grinding.

There is basically little difference between using a grinding tool and a cutting tool to resurface a valve face so the procedure for doing either is the same. Keep in mind that the metal removed from the valve face and valve seat increases the amount of valve stem length on the spring side of the cylinder head when the valve is seated.

Before starting valve seat work, carefully check the seats for cracks. Like valve guides, there are two types of valve seats: integral and insert. Integral seats are part of the casting. Insert seats are pressed into the head and are always used in aluminum cylinder heads.

Cracked integral seats sometimes can be repaired by installing inserts, if the crack is not too deep. Insert seats are pressed into the head and are always used in aluminum cylinder heads. Valve seats can be reconditioned or repaired by one of two methods, depending on the seat type: machining a counterbore to install an insert seat, or grinding, cutting, or machining an integral seat.

When grinding a valve seat, it is very important to select and use the correct size pilot and grind stone. The grinding wheel should be positioned and centered by inserting a properly sized pilot shaft into the valve guide (Figure 10). The seat is ground by continually raising and lowering the grinder unit on and off the seat at a rate of approximately 120 times per minute. Grinding should only continue until the seat is clean and free of defects.

After the seat is ground, valve fit is checked using machinist dye. The valve face should be coated with dye, installed in its seat, and slightly rotated. The valve is then removed and the dye pattern on the valve face and valve seat inspected. If the valve face and seat are not contacting each other evenly,

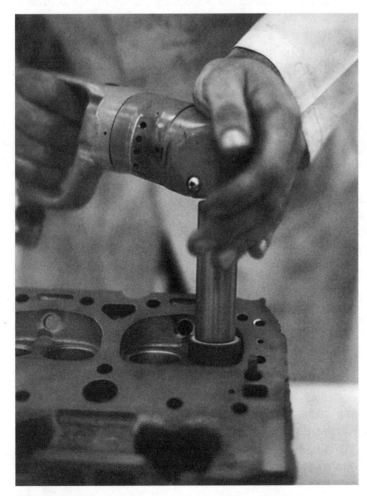

Figure 10 Refacing a valve seat with a seat grinder.

or if the contact line is too high, the seat must be reground with the same stone used initially to correct the condition. If the line is too low or the width is not correct, the seat must be reground with stones of different angles.

Cutting valve seats differs from grinding only in the equipment used. Hardened valve seat cutters replace grinding wheels for seat finishing. The basic seat cutting procedures are the same as those for grinding.

Torque-Indicating Wrench

Torque is the twisting force used to turn a fastener against the friction between the threads and between the head of the fastener and the surface of the component. The fact that practically every vehicle and engine manufacturer publishes a list of torque recommendations is ample proof of the importance of using proper amounts of torque when tightening nuts or bolts. The amount of torque applied to a fastener is measured with a torque-indicating wrench or torque wrench.

There are three basic types of torque-indicating wrenches available with pounds per inch and pounds per foot increments: a beam torque wrench that has a beam that points to the torque reading, a "click"-type torque wrench in which the desired torque reading is set on the handle (when the torque reaches that level, the wrench clicks), and a dial torque wrench that has a dial that indicates the torque exerted on the wrench. Some designs of this type torque wrench have a light or buzzer that turns on when the desired torque is reached.

Torque Angle Gauge

Some manufacturers recommend a torque-angle method for cylinder head bolts, which requires the use of a torque angle gauge. Torque-to-yield bolts must be tightened according to the manufacturer's recommendations. Typically this involves two steps: tighten the bolt to the specified torque, and then tighten the bolt an additional amount. The latter is expressed in degrees. To accurately measure the number of degrees added to the bolt, a torque angle gauge is attached to the wrench. The additional tightening will stretch the bolt, producing a very reliable clamp load that is much higher than can be achieved just by torqueing.

Oil Priming Tool

Prior to starting a freshly rebuilt engine, the oil pump must be primed. There are several ways to prelubricate, or prime, an engine. One method is to drive the oil pump with an electric drill. With most engines, it is possible to make a drive that can be chucked in an electric drill motor to engage the drive on the oil pump. Insert the fabricated oil pump drive extension into the oil pump through the distributor drive hole. To control oil splash, loosely set the valve cover(s) on the engine. After running the oil pump for several minutes, remove the valve cover and see whether there is any oil flow to the rocker arms. If oil reached the cylinder head, the engine's lubrication system is full of oil and is operating properly. If no oil reached the cylinder head, there is a problem with the pump, an alignment of an oil hole in a bearing, or perhaps a plugged gallery.

Using a prelubricator, which consists of an oil reservoir attached to a continuous air supply, is the best method of prelubricating an engine without running it. When the reservoir is attached to the engine and the air pressure is turned on, the prelubricator will supply the engine's lubrication system with oil under pressure.

Cooling System Pressure Tester

A cooling system pressure tester contains a hand pump and a pressure gauge. A hose is connected from the hand pump to a special adapter that fits on the radiator filler neck. This tester is used to pressurize the cooling system and check for coolant leaks. Additional adapters are available to connect the tester to the radiator cap. With the tester connected to the radiator cap, the pressure relief action of the cap may be checked.

Coolant Hydrometer

A coolant hydrometer is used to check the amount of antifreeze in the coolant. This tester contains a pickup hose, coolant reservoir, and squeeze bulb. The pickup hose is placed in the radiator coolant. When the squeeze bulb is squeezed and released, coolant is drawn into the reservoir. As coolant enters the reservoir, a pivoted float moves upward with the coolant level. A pointer on the float indicates the freezing point of the coolant on a scale located on the reservoir housing.

Belt Tension Gauge

A belt tension gauge is used to measure drive belt tension. The belt tension gauge is installed over the belt, and the gauge indicates the amount of belt tension.

Service Manuals

Perhaps the most important tools you will use are service manuals. There is no way a technician can remember all of the procedures and specifications needed to correctly repair all vehicles. Therefore, a good technician relies on service manuals and other information sources for this information. Good information plus knowledge allows a technician to fix a problem with the least bit of frustration and at the lowest cost to the customer.

To obtain the correct engine specifications and other information, you must first identify the engine you are working on. The best source for engine identification is the VIN. The engine code can be interpreted through information given in the service manual. The manual may also help you identify the engine through casting numbers and/or markings on the cylinder block or head.

The primary source of repair and specification information for any car, van, or truck is the manufacturer. The manufacturer publishes service manuals each year, for every vehicle built. Because of the enormous amount of information, some manufacturers publish more than one manual per year per car model. They are typically divided into sections based on the major systems of the vehicle. In the case of engines, there is a section for each engine that may be found in the vehicle. Manufacturers' manuals detail all repairs, adjustments, specifications, diagnostic procedures, and special tools required.

Since many technical changes occur on specific vehicles each year, manufacturers' service manuals need to be constantly updated. Updates are published as service bulletins (often referred to as Technical Service Bulletins or TSBs) that show the changes in specifications and repair procedures during the model year. These changes do not appear in the service manual until the next year. The car manufacturer provides these bulletins to dealers and repair facilities on a regular basis.

Service manuals are also published by independent companies rather than by the manufacturers. However, they pay for and get most of their information from the car makers. They contain component information, diagnostic steps, repair procedures, and specifications for several car makes in one book. Information is usually condensed and is more general in nature than the manufacturer's manuals. The condensed format allows for more coverage in less space and therefore is not always specific. They may also contain several years of models as well as several car makes in one book.

Many of the larger parts manufacturers have excellent guides on the various parts they manufacture or supply. They also provide updated service bulletins on their products. Other sources for up-to-date technical information are trade magazines and trade associations.

The same information that is available in service manuals and bulletins is also available on CD-ROMs and DVDs. A single compact disk can hold 250,000 pages of text. This eliminates the need for a huge library containing all of the printed manuals. Using a CD-ROM to find information is also easier and quicker. The disks are normally updated monthly, and not only contain the most recent service bulletins but also engineering and field service fixes.

CROSS REFERENCE GUIDE

NATEF Task	Job Sheet
A.1	1
A.2	52
A.3	52
A.4	1
A.5	1
A.6	1
A.7	2
A.8	3
A.9	4
A.10	5
A.11	6 & 7
A.12	8 & 9
B.1	10
B.2	11
B.3	12
B.4	13 & 14
B.5	10
B.6	13 (15 & 16 are related)
B.7	17
B.8	18
B.9	19
B.10	19
B.11	20
B.12	21
B.13	22
B.14	23
B.15	24
B.16	25
C.1	53 (26 is related)
C.2	27
C.3	28
C.4	30 (29 is related)
C.5	31
C.6	32
C.7	33
C.8	34
C.9	35
C.10	36

NATEF Task	Job Sheet
C.11	36
C.12	37
C.13	39
C.14	37
C.15	38 (40 is related)
D.1	41
D.2	26
D.3	42
D.4	43
D.5	44
D.6	45
D.7	42
D.8	46
D.9	47
D.10	48
D.11	49
D.12	50
D.13	51

JOB SHEETS

ENGINE REPAIR JOB SHEET 1

Verifying the Condition of and Inspecting an Engine

Name _____ Station _____ Date _____

NATEF Correlation

This Job Sheet addresses the following NATEF tasks:

A.1. Identify and interpret engine concern; determine necessary action.

A.4. Inspect engine assembly for fuel, oil, coolant, and other leaks; determine necessary action.

A.5. Diagnose engine noises and vibrations; determine necessary action.

A.6. Diagnose the cause of excessive oil consumption, unusual engine exhaust color, odor, and sound; determine necessary action.

Objective

Upon completion of this job sheet, you will be able to verify and interpret engine concerns; inspect an engine assembly for fuel, oil, coolant, and other leaks; diagnose engine noises and vibrations; and diagnose the cause of excessive oil consumption, unusual engine exhaust color, odor, and sounds.

Tools and Materials

A road worthy car
Stethoscope

Protective Clothing

Goggles or safety glasses with side shields

Describe the vehicle being worked on:

Year _____ Make _____ Model _____

VIN _____ Engine type and size _____

PROCEDURE

Verify Engine Condition

1. Verifying the customer's concern is typically the first step you should take when diagnosing a problem. If the owner of the vehicle stated a concern, describe it. If there are no customer complaints, describe the general running condition to the best of your knowledge. The concern may be one of performance, smoke, leaks, or noises. In your answer, be sure to completely describe the condition and state when, where, and how the condition occurs.

_____

_____

_____

_____

2. Verify the complaint. Describe what you will do in an attempt to duplicate the concern. If necessary, road test the vehicle under the same conditions that are present when the problem normally occurs. Include in this description conditions that will explain when, where, and how the concern occurs.

_____

_____

_____

_____

3. Often a customer may only notice poor performance during one condition, whereas the problem may exist at others. Also, observing the performance of an engine during a variety of modes of operation may let you know what is working fine and what does not need to be tested further. Describe how the engine performed during the following conditions:

A. Starting:

_____

B. Idling:

_____

C. Slow acceleration:

_____

D. Slow cruise:

_____

E. Slow deceleration:

_____

F. Heavy acceleration:

_____

G. Highway cruise:

_____

H. Fast deceleration:

_____

I. Shut down:

_____

4. Based on the results of the above checks, what engine systems do you think should be checked to find the cause of the customer's complaint?

_____

_____

_____

Engine Leak Diagnosis

1. After you have completed the above exercise, park the vehicle in a place where the floor is fairly clean and dry. Shut the engine off and allow it to cool. Once the engine is cool, look on the shop floor for any evidence of leaks. Remember, engine leaks can cause safety, driveability, and durability concerns. Describe the location and color of the fluid that leaked on the floor.

_____

_____

_____

WARNING: _Gasoline fumes are extremely dangerous! If ignited, they will cause a very serious explosion and fire, resulting in personal injury and property damage. If you suspect that the leak is a fuel leak, immediately inform your instructor of the problem and take all precautions to avoid igniting the fuel._

2. Engine fuel leaks are expensive and dangerous, and they should be corrected immediately when they are detected. If a gasoline odor occurs inside or near a vehicle, it should be inspected for fuel leaks immediately. Inspect the following parts of the vehicle to locate the source of the fuel leak. Describe what you found at each of these components:

A. Fuel tank:

_____

B. Fuel tank filler cap:

_____

C. Fuel lines and filter:

_____

D. Mechanical fuel pump (carbureted engines):

_____

E. Vapor recovery system lines:

_____

F. Carburetor:

_____

G. Pressure regulator, fuel rail, and injectors (fuel injected engines):

_____

3. Based on the above inspection, what do you recommend should be done to the vehicle to correct the fuel leak?

_____

_____

4. Engine oil leaks may cause an engine to run out of oil, resulting in serious engine damage. Sometimes the leak goes unnoticed, and yet the engine's oil needs to be added to more often than normal. A careful inspection may help you locate the source of the leak; however, if it is difficult to locate the exact cause of an oil leak, the engine should be cleaned first. Carefully look at the following engine parts and describe what you see. If oil is found on or around any of these parts, look carefully at the component directly above the leak. Oil runs down the engine surfaces. Also, if an oil leak is found, check the PCV system for blockage and proper operation. A bad PCV system will allow crankcase pressure to build and this may be the ultimate cause of the oil leak.

A. Rear main bearing seal:

_____

B. Expansion plug in rear camshaft bearing:

_____

C. Rear oil gallery plug:

_____

D. Oil pan:

_____

E. Oil filter:

_____

F. Rocker arm covers:

_____

G. Intake manifold front and rear gaskets (V-type engines):

_____

H. Mechanical fuel pump gasket or worn fuel pump pivot pin:

_____

I. Timing gear cover or seal:

_____

J. Front main bearing:

_____

K. Oil pressure sending unit:

_____

L. Distributor O-ring or gasket:

_____

M. Engine casting porosity:

_____

N. Oil cooler or lines (where applicable):

_____

5. Based on the above inspection, what do you recommend should be done to the vehicle to correct the oil leak?

_____

_____

6. If an engine uses excessive oil and there is no evidence of leaks, the oil may be burning in the combustion chambers. If excessive amounts of oil are burned in the combustion chambers, the exhaust contains blue smoke, and the spark plugs may be fouled with oil. Excessive oil burning in the combustion chambers may be caused by worn rings and cylinders or worn valve guides and valve seals. Remove the spark plugs and describe the condition of each.

#1 _____

#2 _____

#3 _____

#4 _____

#5 _____

#6 _____

#7 _____

#8 _____

7. When an engine coolant leak causes low coolant level, the engine quickly overheats and severe engine damage may occur. When you suspect a coolant leak, check the following components and describe what you found:

A. Upper radiator hose:

_____

B. Lower radiator hose:

_____

C. Heater hoses:

_____

D. By-pass hose:

_____

    E. Water pump:

    _____

    F. Engine expansion plugs or block heater:

    _____

    G. Radiator:

    _____

    H. Thermostat housing:

    _____

    I. Heater core:

    _____

8. Based on the above inspection, what do you recommend should be done to the vehicle to correct the oil leak?

    _____

    _____

9. Some coolant leaks are internal to the engine and will not be visible. A whitish exhaust may be indicative of this sort of coolant leak if it is caused by a leaking head gasket or cracked engine components. Take a look at the exhaust while the engine is idling and describe it:

    _____

    _____

10. A cooling system pressure tester is commonly used to locate leaks. The use of this tester is covered in another job sheet in this manual. What page is the procedure found on?

    _____

Engine Exhaust Diagnosis

1. Some engine problems may be diagnosed by the color, smell, or sound of the exhaust. If the engine is operating normally, the exhaust should be colorless. In severely cold weather, it is normal to see a swirl of white vapor coming from the tailpipe, especially when the engine and exhaust system are cold. This vapor is moisture in the exhaust, which is a normal by-product of the combustion process. Carefully observe the exhaust from your test vehicle and describe what you see and/or smell during the following operating conditions:

    A. Cold start-up:

    _____

    B. Cold idle:

    _____

    C. Warm start-up:

    _____

D. Warm idle:

_____

E. Snap-throttle open:

_____

F. Snap-throttle closed:

_____

2. Based on the above, what are your conclusions? (Use the explanations below to guide your thoughts.)

_____

_____

If the exhaust is blue, excessive amounts of oil are entering the combustion chamber, and this oil is being burned with the fuel. When the blue smoke in the exhaust is more noticeable on deceleration, the oil is likely getting past the rings into the cylinder. If the blue smoke appears in the exhaust immediately after a hot engine is restarted, the oil is likely leaking down the valve guides.

- If black smoke appears in the exhaust, the air-fuel mixture is too rich. A restriction in the air intake, such as a plugged air filter, may be responsible for a rich air-fuel mixture.
- Gray smoke in the exhaust may be caused by coolant leaking into the combustion chambers. This may be most noticeable when the engine is first started after it has been shut off for over half an hour.
- On catalytic converter-equipped vehicles, a strong sulfur smell in the exhaust indicates a rich air-fuel mixture. Some sulfur smell on these engines is normal, especially during engine warm-up.

3. You may have noticed a change in sound during the above test. If you did, describe the sound change and operating mode in which the sound changed.

_____

_____

_____

_____

4. Use these guidelines to determine the possible cause of the sound. Then state your best guess for the cause of the noise.

- When the engine is idling, the exhaust at the tailpipe should have a smooth, even sound.
- If, during idle, the exhaust has a "puff" sound at regular intervals, a cylinder may be misfiring.
- When this sound is present, check the engine's ignition and fuel systems, and the engine's compression.
- If the vehicle has excessive exhaust noise, while the engine is accelerated, check the exhaust system for leaks.
- A small exhaust leak may cause a whistling noise when the engine is accelerated.
- If the exhaust system produces a rattling noise when the engine is accelerated, check the muffler and catalytic converter for loose internal components.
- When the engine has a wheezing noise at idle or with the engine running at higher rpm, check for a restricted exhaust system.

_____

_____

_____

Engine Noise Diagnosis

1. Sounds from the engine itself can help you locate engine problems or help you identify a weakness in the engine before it becomes a big problem. Long before a serious engine failure occurs, there are usually warning noises from the engine. Engine defects such as damaged pistons, worn rings, loose piston pins, worn crankshaft bearings, worn camshaft lobes, and loose and worn valve train components usually produce their own peculiar noises. Certain engine defects also cause a noise under specific engine operating conditions. Since it is sometimes difficult to determine the exact location of an engine noise, a stethoscope may be useful. The stethoscope probe is placed on, or near, the suspected component, and the ends of the stethoscope are installed in your ears. The stethoscope amplifies sound to assist in noise location. When the stethoscope probe is moved closer to the source of the noise, the sound is louder in your ears. If a stethoscope is not available, what can be safely used to amplify the sound and help locate the source of the noise? _____

   CAUTION: *When placing a stethoscope probe in various locations on a running engine, be careful not to catch the probe or your hands in moving components such as cooling fan blades and belts.*

2. Since a lack of lubrication is a common cause of engine noise, always check the engine oil level and condition prior to noise diagnosis. Carefully observe the oil for contamination by coolant or gasoline. Check the oil level and condition on your test vehicle, then record your findings:

   _____

   _____

3. During the diagnosis of engine noises, always operate the engine under the same conditions as those that are present when the noise ordinarily occurs. Remember that aluminum engine components such as pistons expand more when heated than cast iron alloy components do. Therefore, a noise caused by a piston defect may occur when the engine is cold but disappear when the engine reaches normal operating temperature. If the customer has an engine noise concern, describe the noise and state when it occurs.

   _____

   _____

   _____

4. Duplicate the condition at which the noise typically occurs and describe all that you hear as you listen to the engine.

   _____

   _____

5. If you verified the customer's concern, use a stethoscope to find the spot where the noise is the loudest. Describe where that is:

   _____

   _____

6. What could be causing the noise to be loud at that spot?

   _____

   _____

7. To help you understand and use noise as a diagnostic tool, you will be given a description of an engine noise. Using your knowledge and any resources (such as your textbook) you have handy, identify the conditions or problems that would cause each of the following noises:

A. A hollow, rapping noise that is most noticeable on acceleration with the engine cold. The noise may disappear when the engine reaches normal operating temperature.

_____

B. A heavy thumping knock for a brief time when the engine is first started after it has been shut off for several hours. This noise may also be noticeable on hard acceleration.

_____

C. A sharp, metallic, rapping noise that occurs with the engine idling.

_____

D. A thumping noise at the back of the engine.

_____

E. A rumbling or thumping noise at the front of the engine, possibly accompanied by engine vibrations. When the engine is accelerated under load, the noise is more noticeable.

_____

F. A light, rapping noise at speeds above 35 mph (21 kph). The noise may vary from a light to a heavier rapping sound depending on the severity of the condition. If the condition is very bad, the noise may be evident when the engine is idling.

_____

G. A high-pitched, clicking noise is noticeable in the upper cylinder area during acceleration.

_____

H. A heavy clicking noise is heard with the engine running at 2,000 to 3,000 rpm. When the condition is severe, a continuous, heavy clicking noise is evident at idle speed.

_____

I. A whirring and light rattling noise when the engine is accelerated and decelerated. Severe cases may cause these noises at idle speed.

_____

J. A light clicking noise with the engine idling. This noise is slower than piston or connecting rod noise and is less noticeable when the engine is accelerated.

_____

K. A high-pitched clicking noise that intensifies when the engine is accelerated.

_____

L. A noise that is similar to marbles rattling inside a metal can. This noise usually occurs when the engine is accelerated.

_____

Instructor's Comments

ENGINE REPAIR JOB SHEET 2

Using a Vacuum Gauge

Name _____ Station _____ Date _____

NATEF Correlation

This Job Sheet addresses the following NATEF task:

A.7. Perform engine vacuum tests; determine necessary action.

Objective

Upon completion of this job sheet, you will have demonstrated the ability to connect and read a vacuum gauge to monitor engine condition.

Tools and Equipment

Vacuum gauge

Various lengths of vacuum hose

Tee hose fittings

Protective Clothing

Safety glasses

Describe the vehicle being worked on:

Year _____ Make _____ Model _____

VIN _____ Engine type and size _____

Describe the general operating condition:

_____

Give a definition of vacuum _____

PROCEDURE

1. Carefully look over the engine's intake manifold to identify vacuum hoses. Select one that is small and easily accessible. DO NOT disconnect it until you have the approval of your instructor. Also make sure the hose is below the throttle plate assembly.   ☐ Task completed

2. With the engine off, disconnect the selected hose. Connect the Tee fitting to the end of the hose and connect another hose from the Tee fitting to the place where the vacuum hose was originally connected.   ☐ Task completed

3. Connect the vacuum gauge to the remaining connection at the Tee fitting.   ☐ Task completed

4. Start the engine and observe the vacuum gauge.   ☐ Task completed

5. Describe the gauge reading and the action of the needle.

   _____

   _____

6. Quickly open the throttle and allow it to quickly close.   ☐ Task completed

7. Describe the gauge reading and the action of the needle.

   _____

   _____

8. Turn off the engine and disconnect the vacuum gauge and hoses. Then   ☐ Task completed
   reconnect the engine's hose to its appropriate connector.

Summarize what you observed.

_____

_____

_____

Instructor's Comments

_____

_____

_____

ENGINE REPAIR JOB SHEET 3

Conduct a Cylinder Power Balance Test

Name _____ Station _____ Date _____

NATEF Correlation

This Job Sheet addresses the following NATEF task:

A.8. Perform cylinder power balance tests; determine necessary action.

Objective

Upon completion of this job sheet, you will be able to conduct a cylinder power balance test and accurately interpret the results.

Tools and Materials

Tune-up scope

Protective Clothing

Goggles or safety glasses with side shields

PROCEDURE

1. Describe the vehicle being worked on.

   Year _____ Make _____ VIN _____

   Model _____

   Engine type and size _____ Firing order _____

2. Describe the general running condition of the engine.

   _____

   _____

   _____

3. Connect the scope to the engine according to the instructions given with ☐ Task completed
   the equipment.

4. Conduct a cylinder power balance test on the engine, using proper testing procedures (make sure not to short a cylinder for a long time if the vehicle is equipped with a catalytic converter), then record the results below:

Cylinder #	1	2	3	4	5	6	7	8
RPM loss	___	___	___	___	___	___	___	___

5. Describe what is indicated by the results of this test.

_____

_____

_____

6. Briefly explain what is actually being measured by a cylinder power balance test.

_____

_____

7. Based on the test results, describe the condition of the engine. (Does it agree with your original description of the engine?)

_____

_____

_____

8. How and why would the readings be different if the camshaft intake lobes for the number two cylinder were severely worn?

_____

_____

_____

Instructor's Comments

_____

_____

_____

ENGINE REPAIR JOB SHEET 4

Perform a Cylinder Compression Test

Name _____ Station _____ Date _____

NATEF Correlation

This Job Sheet addresses the following NATEF task:

A.9. Perform cylinder compression tests; determine necessary action.

Objective

Upon completion of this job sheet, you will be able to perform cylinder compression tests.

Tools and Materials

Spark plug socket, extensions, and ratchet   Air blower
Compression tester   Small oil can with flexible nozzle
Fender covers

Protective Clothing

Safety goggles or glasses with side shields

Describe the vehicle being worked on:

Year _____ Make _____ Model _____

VIN _____ Engine type and size _____

Describe general condition:

_____

PROCEDURE

1. Disable the ignition or fuel injection system.   ☐ Task completed

2. Remove the spark plug cables from all spark plugs. Use an air blower to   ☐ Task completed
   clean around the spark plugs. Remove all of the spark plugs and place them
   on a clean surface in the order in which they are removed from the engine.

3. Remove the air cleaner from the car. Block the throttle in its wide-open   ☐ Task completed
   position.

4. Install the compression testing adapter (if necessary) into the number 1   ☐ Task completed
   spark plug hole, and connect the compression tester to it.

5. Crank the engine through four compression strokes and record the read-   ☐ Task completed
   ing on the compression gauge.

6. Remove the compression adapter from the spark plug hole. Squirt a small   ☐ Task completed
   amount of oil into the cylinder through the spark plug hole.

7. Repeat steps 4 and 5 on the same cylinder.   ☐ Task completed

8. Complete steps 4, 5, 6, and 7 for each of the remaining cylinders in the   ☐ Task completed
   engine.

9. Clean each spark plug and replace each spark plug in the cylinder from   ☐ Task completed
   where it was removed. Reconnect the spark plug cables. Remove whatev-
   er tool is blocking the throttle plate open, and reinstall the air cleaner.

10. Start the engine and check its operation to determine if everything was con- ☐ Task completed
    nected back properly.

Problems Encountered

_____

_____

_____

Instructor's Comments

_____

_____

_____

REPORT SHEET ON COMPRESSION TEST		
Compression for Each Cylinder _Cylinder No._	_Dry_	_Wet_
1.		
2.		
3.		
4.		
5.		
6.		
7.		
8.		

Conclusions and Recommendations _____

_____

_____

ENGINE REPAIR JOB SHEET 5

Perform a Cylinder Leakage Test

Name _____ Station _____ Date _____

NATEF Correlation

This Job Sheet addresses the following NATEF task:

A.10. Perform cylinder leakage tests; determine necessary action.

Objective

Upon completion of this job sheet, you will be able to conduct cylinder leakage tests.

Tools and Materials

Breakover bar	Radiator coolant (if applicable)
Chalk	Screwdrivers
Compressed air	Service manual
Fender covers	Socket set
Indicator light	Spark plug socket and ratchet
Jumper lead	TDC indicator
Leakage tester and whistle	Test adapter hose

Protective Clothing

Safety goggles or glasses with side shields

Describe the vehicle being worked on:

Year _____ Make _____ Model _____

VIN _____ Engine type and size _____

Describe general condition:

_____

PROCEDURE

CAUTION: *Very high voltages are present with high-energy ignition systems. Do NOT use this procedure on cars with Distributorless ignition systems unless the ignition system is totally disabled.*

1. Check the coolant level and fill, if needed. Run the engine until it reaches normal operating temperature. Then, turn it off and disable the ignition or fuel injection system. ☐ Task completed

2. Disconnect the spark plug cables from the plugs. Use compressed air to clean all foreign matter out of the plug wells. Remove all spark plugs. Set the plugs on a workbench or other clean surface in the order in which they were removed. Remove all plug gaskets or tubes from the cylinder head. ☐ Task completed

3. Remove the air cleaner. Block the throttle plate in a wide-open position □ Task completed
   using a screwdriver or similar tool. Disconnect the PCV hose from the
   crankcase.

4. Install the test adapter hose in the number 1 cylinder spark plug hole. Con- □ Task completed
   nect the tester whistle to adapter hose.

5. Using a socket and ratchet on the crankshaft pulley nut or bolt, slowly rotate □ Task completed
   the engine in its normal direction until the test whistle sounds indicating
   the beginning of the compression stroke. Continue rotation until the timing mark on the crankshaft pulley lines up with the engine-timing pointer
   on the timing chain cover. Remove the tester whistle from the adapter hose.

6. Connect the tester to the adapter hose. Does the gauge show more than 20 □ Task completed
   percent leakage? If so, look for air leaking from the throttle plate, tailpipe,
   or crankcase and for air bubbles in the radiator. Record your results on the
   Report Sheet on Cylinder Leakage Test.

7. Disconnect the tester from the adapter hose. Resume rotating engine using □ Task completed
   socket and ratchet on the crankshaft nut or bolt until the next appropriate
   cylinder mark lines up with the chalk mark on the engine. It may be necessary to use the whistle to locate the TDC.

8. Remove the adapter from the previously tested cylinder and install it in the □ Task completed
   plug hole of the next cylinder in the firing order.

9. Repeat steps 6, 7, and 8 on each remaining cylinder. □ Task completed

Problems Encountered

_____

_____

_____

Instructor's Comments

_____

_____

_____

Name _____   Station _____   Date _____

	REPORT SHEET ON CYLINDER LEAKAGE TEST	
No.	*Percentage*	*Leakage From*
1.		
2.		
3.		
4.		
5.		
6.		
7.		
8.		

Conclusions and Recommendations _____

_____

_____

ENGINE REPAIR JOB SHEET 6

Removing an Engine from a FWD Vehicle

Name _____ Station _____ Date _____

NATEF Correlation

This Job Sheet addresses the following NATEF task:

A.11. Remove and reinstall engine in a late model front-wheel-drive vehicle (OBD-I or newer); reconnect all attaching components and restore the vehicle to running condition.

Objective

Upon completion of this job sheet, you will be able to remove the engine from a front-wheel-drive vehicle.

Tools and Materials
Engine cradle and dolly

Drain pans

Battery terminal puller

Fender covers

Hoist

Portable crane

Hand tools

Protective Clothing
Goggles or safety glasses with side shields

Describe the vehicle being worked on:
Year _____ Make _____ Model _____

VIN _____ Engine type and size _____

PROCEDURE

1. Center the vehicle on a frame contact lift/hoist.   ☐ Task completed

2. Go through the vehicle and check for any memory functions for the radio, power seats, etc. Write down all settings so they can be reset after the engine is reinstalled. What memory devices are on this vehicle?

   _____

   _____

3. Remove the battery. Which cable did you disconnect first? Why?

   _____

   _____

4. If the hood must be removed, mark the location of the hinge to the hood for reference during assembly, then remove the hood and place it in a safe location. Where did you set it?

_____

_____

5. Drain the engine oil.   □ Task completed

6. Drain the engine's coolant.   □ Task completed

7. According to the service manual, will the transaxle be removed with the engine? If so, drain its fluid.

_____

_____

8. Disconnect the air induction system and remove the air cleaner assembly.   □ Task completed

9. Relieve fuel pressure at the fuel rail. How did you do that?

_____

_____

10. Disconnect the fuel supply line to the fuel rail and the fuel return line to   □ Task completed
the pressure regulator.

11. Disconnect the throttle linkage at the throttle body. Are there brackets for the linkage that need to be removed so the linkage can be moved out of the way?

_____

_____

12. Remove all drive belts.   □ Task completed

13. Attempt to remove the A/C compressor without disconnecting the refrigerant lines. Then, move it out of the way. If you need to disconnect the lines, what do you need to do before doing this? And, what should you do after the lines are disconnected?

_____

_____

14. Remove the power steering lines to the pump.   □ Task completed

15. Remove any other hoses or lines that are connected to the engine.   □ Task completed

16. Disconnect the heater inlet and outlet hoses.   □ Task completed

17. Disconnect the electric cooling fans.   □ Task completed

18. Remove the radiator; make sure all electrical connections to the radiator are disconnected first. Were there electrical connections?

_____

_____

19. Label and disconnect all remaining wires and vacuum hoses.   □ Task completed

20. Make sure the engine ground strap is disconnected.   □ Task completed

21. Loosen and remove the axle shaft hub nuts.   □ Task completed

22. Raise the vehicle so you can comfortably work under the vehicle.   □ Task completed

23. Remove the wheel and tire assemblies for the front wheels.   □ Task completed

24. Disconnect all suspension and steering parts that need to be removed according to the service manual. Index the parts so wheel alignment will be close after reassembly. What needed to be removed?

    _____

    _____

25. Remove the axle shafts.   □ Task completed

26. Disconnect the fuel line from the fuel tank to the engine. What did you use to plug the line and prevent fuel leakage?

    _____

    _____

27. Disconnect all lines, hoses, and electrical wiring to the transmission, if the transmission is to be removed with the engine. What did you need to disconnect?

    _____

    _____

28. Disconnect all linkages to the transmission; again, do this only if the transmission comes out with the engine. What linkages were there?

    _____

    _____

29. Remove any heat shields that may be in way.   □ Task completed

30. Disconnect the exhaust pipes at the exhaust manifold.   □ Task completed

If the engine is to be removed from the top of the engine compartment:

1. Connect the engine sling or lifting chains to the engine. Where did you attach the chains?

    _____

    _____

2. Connect the sling to the crane and raise the crane just enough to support   □ Task completed
   the engine.

3. From under the vehicle, remove the cross member.   □ Task completed

4. Remove the mounting bolts for the engine at the engine and transmission   □ Task completed
   mounts.

5. From under the hood, remove all remaining mounts.   □ Task completed

6. Raise the engine slightly to free the engine from the mounts. ☐ Task completed

7. Slowly raise the engine from the engine compartment. Guide the engine around all wires and hoses to make sure nothing gets damaged. Did you have any problems doing this?

_____

_____

8. Once the engine is cleared from the vehicle, prepare to separate it from the ☐ Task completed
   transaxle and mount it on an engine stand.

If the engine is to be removed from the bottom of the vehicle:

1. Position the engine cradle and dolly under the engine. ☐ Task completed

2. Adjust the pegs of the cradle so they fit into the recesses on the bottom of ☐ Task completed
   the engine and secure the engine.

3. Remove all engine and transmission mount bolts. ☐ Task completed

4. Slowly raise the vehicle, lifting it slightly away from the engine. ☐ Task completed

5. Check the area around the engine to make sure nothing remains that should ☐ Task completed
   be removed or disconnected.

6. Slowly raise the vehicle over the engine while guiding all wires and hoses out of the way. Did you have any problems doing this?

_____

_____

7. Once the vehicle is clear of the engine, prepare to separate the engine from ☐ Task completed
   the transaxle and mount it on an engine stand.

Instructor's Comments

_____

_____

_____

ENGINE REPAIR JOB SHEET 7

Installing an Engine into a FWD Vehicle

Name _____ Station _____ Date _____

NATEF Correlation

This Job Sheet addresses the following NATEF task:

A.11. Remove and reinstall engine in a late model front-wheel-drive vehicle (OBD-I or newer); reconnect all attaching components and restore the vehicle to running condition.

Objective

Upon completion of this job sheet, you will be able to reinstall the engine in a front-wheel drive vehicle.

Tools and Materials

Engine cradle and dolly

Hoist

Portable crane

Hand tools

Protective Clothing

Goggles or safety glasses with side shields

Describe the vehicle being worked on:

Year _____ Make _____ Model _____

VIN _____ Engine type and size _____

PROCEDURE

1. Was the engine removed from the top or underneath? _____

If the engine will be installed through the top:

2. Connect the engine to a sling and then connect the sling to the crane. What did you connect the sling to?

   _____

   _____

3. Slowly lower the engine into the engine compartment. Guide the engine around all wires and hoses to make sure nothing gets damaged. Did you have any problems doing this?

   _____

   _____

4. As the engine approaches its position in the engine compartment, align the engine and transmission mounts. Did you have any problems doing this?

_____

_____

5. Once the mounts are aligned, lower the engine so you can install the bolts into the mounts. ☐ Task completed

6. Raise the vehicle to a good working height. ☐ Task completed

If the engine will be installed from under the car:

2. Install the engine onto the engine cradle and dolly. What did you use to secure the engine onto the dolly?

_____

_____

3. Lift the vehicle on a hoist or lift. ☐ Task completed

4. Position the engine under the vehicle. ☐ Task completed

5. Slowly lower the vehicle over the engine while guiding all wires and hoses out of the way. Did you have any problems doing this?

_____

_____

6. As the vehicle gets close to the engine, align the engine and transmission mounts. Did you have any problems doing this?

_____

_____

7. Once the mounts are aligned, lower the vehicle so you can install the bolts into the mounts. ☐ Task completed

8. Raise the vehicle to a good working height. ☐ Task completed

After the engine is resting on its mounts:

1. While working under the vehicle, install the axle shafts. ☐ Task completed

2. Install the remaining engine and transaxle mounts and braces. What do these connect to?

_____

_____

_____

3. Connect the exhaust manifold to the exhaust system. Was a new gasket required? Did you have any problems doing this?

_____

_____

4. Install any heat shields that were removed when the engine was removed.   ☐ Task completed

5. Connect all linkages to the transmission. What linkages were there?

_____

_____

6. Connect all lines, hoses, and electrical wiring to the transmission. What did you need to connect?

_____

_____

7. Reconnect all suspension and steering parts that were disconnected or removed. After you do this, will the vehicle need a front wheel alignment? Why?

_____

_____

8. Install the wheels and tires. What is the torque spec for the wheel lugs?

_____

9. Tighten the axle hub nuts. What is the torque spec for these nuts?

_____

10. Reconnect the fuel line from the fuel tank to the engine. Where is this connection made?

_____

_____

11. Lower the vehicle so you can work under the hood.   ☐ Task completed

12. Connect any remaining disconnected fuel lines.   ☐ Task completed

13. Connect heating system hoses. Did you need to replace any? Why?

_____

_____

14. Connect the engine ground strap.   ☐ Task completed

15. Connect all electrical connectors and wires.   ☐ Task completed

16. Connect all vacuum and other hoses.   ☐ Task completed

17. Connect the throttle linkage and adjust it according to manufacturer's procedures. Did you need to adjust the linkage? If so, how did you do it?

_____

_____

18. Install the radiator and the cooling fan(s).   ☐ Task completed

19. Connect the rest of the hoses for the cooling system. Did you need to replace any of these? Why?

_____

_____

20. Install the air induction system.   ☐ Task completed

21. Connect any remaining items.   ☐ Task completed

22. Install the battery and connect the battery cables. Which cable did you connect first? Why?

_____

_____

_____

_____

23. Fill the radiator with coolant and visually check for leaks. Did you find any?

_____

_____

24. If the engine doesn't have oil in it already, add the specified amount of the   ☐ Task completed
    proper type of oil.

25. Prime the oil pump of the engine. How did you do that?

_____

_____

26. Prepare the engine for startup. What did you need to do?

_____

_____

Instructor's Comments

_____

_____

_____

ENGINE REPAIR JOB SHEET 8

Removing an Engine from a RWD Vehicle

Name _____ Station _____ Date _____

NATEF Correlation

This Job Sheet addresses the following NATEF task:

A.12. Remove and reinstall engine in a late model rear-wheel-drive vehicle (OBD-I or newer); reconnect all attaching components and restore the vehicle to running condition.

Objective

Upon completion of this job sheet, you will be able to remove the engine from a rear-wheel drive vehicle.

Tools and Materials

Drain pans	Plugs
Fender covers	Screwdrivers
Lift	Service manual
Masking tape	Sockets and ratchet
Penetrating oil	Transmission jack
Pliers	Wrenches

Protective Clothing

Safety goggles or glasses with side shields

Describe the vehicle being worked on:

Year _____ Make _____ Model _____

VIN _____ Engine type and size _____

Describe general condition:

_____

WARNING: *The technician should always follow the specific engine removal and disassembly procedures given in the service manual for the particular vehicle being worked on. If the following procedure differs at any point from the service manual procedure, follow the procedure given in the service manual.*

PROCEDURE

1. Drive the vehicle onto the lift. Open the hood and install fender covers on the fenders. Mask any areas where there is any possibility of scratching the paint.   ☐ Task completed

2. Scribe a line to mark the location of the hood hinges for reference in reassembly. Unbolt the hood hinges and remove the hood.   ☐ Task completed

3. Disconnect the battery cables from battery. ☐ Task completed

   CAUTION: *Always disconnect the negative side of the battery first.*

   WARNING: *Always wear eye protection when working around the battery. Any type of spark can cause the battery to explode, resulting in severe injury.*

4. Remove the battery from the vehicle and store it so that it is out of the way ☐ Task completed
   and in a place where it is unlikely that something can easily short across
   the negative and positive battery terminals. It is wise to slow charge the
   battery while it is sitting.

5. Disconnect the ground cable from the cylinder head or engine block. On ☐ Task completed
   some engines it will be necessary to remove the separate ground strap from
   the engine block to the bulkhead or from the engine block to the chassis
   frame near the engine mounts.

6. Remove the air cleaner, intake air duct, and heat tube to aid visibility and ☐ Task completed
   to increase the size of the working area.

7. Drain the cooling system by removing the radiator cap, then opening the ☐ Task completed
   petcock near the bottom of the radiator.

   WARNING: *Do not open the radiator cap or attempt to drain the radiator if the engine is still hot. Allow the engine to cool before opening the cap or severe burns may result.*

8. Disconnect the engine coolant hoses at the top and bottom of the radiator. ☐ Task completed
   If the vehicle is equipped with an automatic transmission, disconnect and
   plug the oil cooler lines at the radiator.

9. Loosen the belt that drives the water pump. Remove the fan shroud, fan ☐ Task completed
   assembly, and water pump pulley. Remove the radiator.

10. Disconnect the heater hoses from the water pump and engine block or ☐ Task completed
    intake manifold.

11. Disconnect and plug all fuel lines to prevent fuel loss and keep dirt from ☐ Task completed
    entering the system. Disconnect the throttle linkage from the throttle body.

    CAUTION: *Most fuel injection lines are under pressure even when the engine is off. Check the bleed-down procedure in the service manual before disconnecting a fuel injection line.*

    WARNING: *Wear eye protection to prevent gasoline, under pressure, from causing irritation or blindness. Also, do not allow gasoline to come into contact with any hot engine components.*

12. Remove the throttle body assembly from the intake manifold. ☐ Task completed

13. Remove the carbon canister, and any fuel system or emissions-related com- ☐ Task completed
    ponent that could pose engine removal problems.

14. Before disconnecting any vacuum lines or electrical lines, attach a piece of masking tape to both sides of the parts that are to be disconnected. Mark the same code letter or number on both ends of the connector.   ☐ Task completed

15. Disconnect all vacuum hoses, wiring connections, and harnesses. On models equipped with electronic or computer controls, disconnect the air charge temperature sensor, engine coolant temperature sensor, and exhaust gas oxygen sensor electrical connectors. Disconnect the ignition coil, water temperature sending unit, and oil pressure sending unit.   ☐ Task completed

16. Remove the generator, distributor, distributor cap, and spark plug wires from the engine. If the vehicle is equipped with power steering, remove the drive belt. Remove the bolts that attach the pump bracket to the engine block and set the assembly aside in the engine compartment with the hoses connected. Secure the pump with wire or twine.   ☐ Task completed

17. On models with air conditioners, remove the compressor mounting bracket attaching bolts. Remove the compressor and mounting bracket assembly and set them aside without disconnecting the refrigerant lines.   ☐ Task completed

    CAUTION: *If the lines are disconnected, the system will require recharging during reassembly.*

    WARNING: *If the air conditioner lines must be disconnected, use only an EPA-approved recovery system to evacuate the refrigerant. Always wear eye protection when servicing the air conditioning system.*

    WARNING: *Improper evacuation procedures can result in frostbite or asphyxiation.*

18. Raise the vehicle following basic safety procedures. Drain the engine's oil. Disconnect the speedometer cable where it attaches to the transmission.   ☐ Task completed

19. Disconnect the exhaust system from the manifold. Loosen the exhaust pipe clamp and slide off the support bracket on the engine. To help in removing bolts and studs, soak them with penetrating oil.   ☐ Task completed

20. If the engine is equipped with a turbocharger, remove it.   ☐ Task completed

21. Remove the flywheel or converter housing cover. If equipped with a manual transmission, remove the bellhousing lower attaching bolts. If equipped with an with automatic transmission, remove the converter-to-flywheel bolts, then remove the converter housing attaching bolts.   ☐ Task completed

22. Remove the starter motor and all necessary brake system components.   ☐ Task completed

23. Attach an engine sling to the engine and support the weight of the engine with a hoist. Then disconnect the engine mounts from the brackets on the frame. Remove all bolts attaching the bellhousing to the engine block. If so equipped, disconnect the clutch linkage.   ☐ Task completed

24. Lower the vehicle and position a suitable transmission jack under the transmission. Raise the jack enough to support the transmission. It may be necessary to raise the transmission while removing the engine.   ☐ Task completed

25. Make a final check to ensure that everything has been disconnected from the engine. The engine is now ready to be lifted from its compartment.   ☐ Task completed

26. Using the Report Sheet on Engine Removal, carefully examine the condition of all parts and indicate your findings.   ☐ Task completed

Problems Encountered

_____

_____

_____

Instructor's Comments

_____

_____

_____

Name _____ Station _____ Date _____

REPORT SHEET ON ENGINE REMOVAL			
Component	*OK*	*Repair*	*Replace*
1. Radiator			
2. Radiator hoses			
3. Heater hoses			
4. Hose clamps			
5. Battery cables			
6. Engine mounts			
7. Transmission mounts			
8. Belts (all)			
9. Water pump			
10. Throttle plate assembly			
11. Clutch and/or shift linkage			
12. Wire connections			
13. Air filter			
14. Oil filter			
15. Vacuum lines			
16. Oil filler cap			
17. PCV valve hoses			
18. Dipstick			
19. Antifreeze			
20. Other items			

ENGINE REPAIR JOB SHEET 9

Installing an Engine in a RWD Vehicle

Name _____ Station _____ Date _____

NATEF Correlation

This Job Sheet addresses the following NATEF task:

A.12. Remove and reinstall engine in a late model rear-wheel-drive vehicle (OBD-I or newer); reconnect all attaching components and restore the vehicle to running condition.

Objective

Upon completion of this job sheet, you will be able to reinstall the engine in a rear-wheel drive vehicle.

Tools and Materials

Fender covers

Lift or hoist

Portable crane

Engine sling or chains

Service manual

Hand tools

Protective Clothing

Goggles or safety glasses with side shields

Describe the vehicle being worked on:

Year _____ Make _____ Model _____

VIN _____ Engine type and size _____

PROCEDURE

1. Connect the engine to a sling and then connect the sling to the crane. What did you connect the sling to?

   _____

   _____

2. Place a transmission jack under the transmission to hold it in position.   ☐ Task completed

3. Slowly lower the engine into the engine compartment. Guide the engine around all wires and hoses to make sure nothing gets damaged. Did you have any problems doing this?

   _____

   _____

4. As the engine approaches its position in the engine compartment, align the engine to the input shaft of the transmission or the torque converter hub into the front pump. Did you have any problems doing this?

_____

_____

5. Carefully wiggle the engine until the input shaft slides through clutch disc splines or the torque converter seats fully into the transmission.   ☐ Task completed

6. Install and tighten the transmission to engine bolts.   ☐ Task completed

7. Start the engine mount bolts into their bores; you may need to wiggle the engine some to do this.   ☐ Task completed

8. Once the mount bolts are in place, tighten them and remove the transmission jack and engine sling.   ☐ Task completed

9. Raise the vehicle to a good working height.   ☐ Task completed

10. While working under the vehicle, install all remaining engine and transmission mounts.   ☐ Task completed

11. Connect the exhaust manifold to the exhaust system. Was a new gasket required? Did you have any problems doing this?

_____

_____

_____

12. Install any heat shields that were removed when the engine was removed.   ☐ Task completed

13. Reconnect the fuel line from the fuel tank to the engine. Where is this connection made?

_____

_____

14. Lower the vehicle so you can work under the hood.   ☐ Task completed

15. Connect any remaining disconnected fuel lines.   ☐ Task completed

16. Connect heating system hoses. Did you need to replace any? Why?

_____

_____

17. Connect the engine ground strap.   ☐ Task completed

18. Connect all electrical connectors and wires.   ☐ Task completed

19. Connect all vacuum and other hoses.   ☐ Task completed

20. Connect the throttle linkage and adjust it according to the manufacturer's procedures. Did you need to adjust the linkage? If so, how did you do it?

_____

_____

21. Install the radiator and the cooling fan(s). ☐ Task completed

22. Connect the rest of the hoses for the cooling system. Did you need to replace any of these? Why?

_____

_____

23. Install the air induction system. ☐ Task completed

24. Connect any remaining items. ☐ Task completed

25. Install the battery and connect the battery cables. Which cable did you connect first? Why?

_____

_____

26. Fill the radiator with coolant and visually check for leaks. Did you find any?

_____

_____

27. If the engine doesn't have oil in it already, add the specified amount of the proper type of oil. ☐ Task completed

28. Prime the oil pump of the engine. How did you do that?

_____

_____

29. Prepare the engine for startup. What did you need to do?

_____

_____

30. Install and align the hood. ☐ Task completed

Instructor's Comments

_____

_____

_____

ENGINE REPAIR JOB SHEET 11

Install Cylinder Heads and Gaskets

Name _____ Station _____ Date _____

NATEF Correlation

This Job Sheet addresses the following NATEF task:

B.2. Install cylinder heads and gaskets; tighten according to manufacturer's specifications and procedures.

Objective

Upon completion of this job sheet, you will be able to install a cylinder head, cylinder head gasket, and tighten it according to the manufacturer's specifications.

Tools and Materials

Basic hand tools

Service manual

Torque wrench

Protective Clothing

Goggles or safety glasses with side shields

Describe the vehicle being worked on:

Year _____ Make _____ Model _____

VIN _____ Engine type and size _____

PROCEDURE

1. Make sure the sealing surfaces on the engine block and cylinder head are clean. Wipe them down with a clean rag prior to installing the cylinder head. ☐ Task completed

2. Carefully look over the head bolts. Are they torque-to-yield bolts? What difference does that make during installation?

   _____

   _____

3. Inspect the threads of the bolts and replace any that need to be replaced. Make sure the replacement bolts are an exact duplicate of what was originally used. Your findings:

   _____

   _____

4. Sort the head bolts and identify any that are longer than the others. Are all of your head bolts the same length?

_____

5. Thoroughly clean the threads of each bolt. Does the service manual recommend that the bolts be put in dry or should they have a lubricant? If yes, what?

_____

6. If some bolts are longer than others, check the service manual to determine their proper location. Where do the longer head bolts go in this engine?

_____

_____

7. Place the cylinder head gasket onto the block and check its fit. Make sure all bores and passages line up properly and that the correct side of the gasket is facing up.   ☐ Task completed

8. Carefully lower the cylinder head over the head gasket and onto the block. Were there dowels to help position the head or did you need to guess at the proper position?

_____

_____

9. Insert the cylinder head bolts into their proper bore.   ☐ Task completed

10. Start each of the bolts by hand-turning them.   ☐ Task completed

11. Refer to the service manual for the proper tightening sequence for the head bolts. Describe that sequence here:

_____

_____

12. Check the specifications for tightening steps. If the manufacturer recommends steps, describe them here.

_____

_____

13. Tighten the bolts according to the service manual's specifications. The final amount of torque on the head bolts should be: _____

Problems Encountered

_____

_____

_____

Instructor's Comments

_____

_____

_____

ENGINE REPAIR JOB SHEET 14

Replace Valve Stem Seals, in Vehicle

Name _____ Station _____ Date _____

NATEF Correlation

This Job Sheet addresses the following NATEF task:

B.4. Replace valve stem seals on an assembled engine; inspect valve retainers, locks, and valve grooves; determine necessary action.

Objective

Upon completion of this job sheet, you will be able to replace valve stem seals with the engine in a vehicle.

Tools and Materials

Hand tools

Spark plug socket

Adapter to put compressed air into the cylinders

Valve spring compressor

Small magnet

ID tags or tape

Service manual

Protective Clothing

Goggles or safety glasses with side shields

Describe the engine being worked on:

Year _____ Make _____ Model _____

VIN _____ Engine type and size _____

Describe general condition:

_____

PROCEDURE

1. Remove the spark plugs from the engine and store them in the order they were in the engine. ☐ Task completed

2. Look over the engine and determine what electrical connectors and hoses must be disconnected in order to gain access to the valve cover(s). Then disconnect those that need to be disconnected. ☐ Task completed

3. Remove the valve covers. ☐ Task completed

4. Remove the rocker arms or cam followers, keeping them in the order in which they are removed.   ☐ Task completed

5. Install the compressed air adapter into the first cylinder spark plug hole.   ☐ Task completed

6. Rotate the engine to top dead center on the same cylinder that you installed the adapter.   ☐ Task completed

7. Hook a compressed air hose to the adapter.   ☐ Task completed

8. With compressed air holding the valves closed, use the valve spring compressor to compress the valve spring. Remove the retainer locks with the magnet.   ☐ Task completed

9. Remove the retainer, spring, and old valve stem seal.   ☐ Task completed

10. Install the new valve stem seal, being careful not to cut the seal on the end of the valve stem.   ☐ Task completed

11. Reverse the removal procedure to reinstall the valve spring, retainer, and locks.   ☐ Task completed

12. Repeat the preceding procedures for the valves on each cylinder. Remember to always put compressed air in the cylinders before removing the locks.   ☐ Task completed

13. When all of the valve stem seals are replaced, reinstall the rocker arms or cam followers.   ☐ Task completed

14. Clean, gap, and replace all of the spark plugs.   ☐ Task completed

15. Adjust the valves to the manufacturer's specifications. Then install the valve cover(s).   ☐ Task completed

16. Reinstall all of the vacuum hoses and the wires that were previously disconnected.   ☐ Task completed

Problems Encountered

_____

_____

_____

Instructor's Comments

_____

_____

_____

ENGINE REPAIR JOB SHEET 15

Recondition Valve Faces

Name _____ Station _____ Date _____

NATEF Correlation

This Job Sheet is related to the following NATEF task:

B.6. Inspect valves and valve seats; determine necessary action.

Objective

Upon completion of this job sheet, you will be able to resurface the face of valves.

Tools and Materials

Measuring scale

Service manual

Valve grinder

Protective Clothing

Goggles or safety glasses with side shields

Describe the engine being worked on:

Year _____ Make _____ Model _____

VIN _____ Engine type and size _____

Describe general condition:

_____

PROCEDURE

1. Locate the following specifications in the service manual for this engine   ☐ Task completed
   and record them:

   Intake valve face angle   _____

   Exhaust valve face angle   _____

   Intake valve minimum margin   _____

   Exhaust valve minimum margin   _____

2. Mount the stem of one of the valves in the V-bracket of the valve grinder.   ☐ Task completed
   Turn on the machine and adjust the coolant flow over the grinding wheel.

   WARNING: *Always wear safety goggles or glasses with side shields when completing this task.*

3. Advance the valve stem toward the wheel and grind just enough material   ☐ Task completed
   off the stem to resurface it.

4. Install the valve stem in the fixture and chamfer the tip.   ☐ Task completed

5. Repeat the stemming operation for each valve.   ☐ Task completed

6. Adjust the valve grinding chuck to the correct angle to grind the valves.   ☐ Task completed
   Inspect the valve-face grinding stone. If it appears to require truing, check
   with your instructor.

7. Adjust the chuck sleeve and chuck stop to accept the valve stems. Install   ☐ Task completed
   a valve in the chuck and adjust the carriage plate stop to prevent the valve
   neck from contacting the grinding wheel.

8. Turn on the grinder and adjust the coolant flow over the grinding wheel.   ☐ Task completed
   Move the valve into grinding position and slowly bring the wheel into contact with it. Move the valve back and forth over the grinding wheel. Be sure
   not to let the valve face pass beyond the edge of the grinding wheel.

9. Move the grinding wheel toward the valve in small increments until the   ☐ Task completed
   face is smooth and shiny all the way around. Be careful to remove the minimum amount of metal. Back the grinding wheel away from the valve and
   then move the valve carriage back, out of the way.

10. Use a 1/64" measuring scale to determine whether the margin is still an   ☐ Task completed
    acceptable width.

11. Grind each of the other valves and record the margin measurements in the   ☐ Task completed
    spaces below. Valves with thin margins must be replaced.

Cylinder Number	Intake	Exhaust
1	_____	_____
2	_____	_____
3	_____	_____
4	_____	_____
5	_____	_____
6	_____	_____
7	_____	_____
8	_____	_____

Conclusions and Recommendations

_____

_____

_____

Problems Encountered

_____

_____

_____

Instructor's Comments

_____

_____

_____

ENGINE REPAIR JOB SHEET 16

Resurface Valve Seats

Name _____ Station _____ Date _____

NATEF Correlation

This Job Sheet is related to the following NATEF task:

B.6. Inspect valves and valve seats; determine necessary action.

Objective

Upon completion of this job sheet, you will be able to resurface valve seats using a valve seat grinder.

Tools and Materials

Service manual
Valve seat grinder with assorted grinding stones and pilots
Cutting oil
Stone dressing tool

Protective Clothing

Goggles or safety glasses with side shields

Describe the vehicle being worked on:

Year _____ Make _____ Model _____

VIN _____ Engine type and size _____

PROCEDURE

1. Carefully inspect the seats in the cylinder head for cracks. Record your findings:

   _____

   _____

2. Check the seats for looseness. Describe how you did this and what your findings were:

   _____

   _____

   _____

3. Clean the seats with emery cloth to remove any residue.   ☐ Task completed

4. Refer to the service manual and find the inside diameter of the valve guides.
   The specification is: _____

5. Select the properly sized pilot for the guides. What is the size of the pilot? Do all of the guides require the same size pilot?

_____

_____

6. Refer to the service manual for the correct seat face angle. The specified angle is:

_____

7. Determine if the seats are hard or soft. If they are hard, use a soft stone. If they are soft (such as cast iron), use a hard stone. Select a grinding stone with the proper texture and size and with the correct angle for the valve seats. Do the intake valve seats require a different size stone than the exhaust valve seats?

_____

_____

8. Check the surface of the stone. Is it flat and clean? Does it need to be dressed? If the stone needs to be dressed, do so now.

_____

_____

9. Insert the pilot into the guide of the seat to be ground. Put a little cutting oil on the seat.   ☐ Task completed

10. Place the grinding stone over the pilot. Make sure the stone doesn't contact any other part of the combustion chamber.   ☐ Task completed

11. Attach the grinding machine to the end of the stone and turn the grinder on.   ☐ Task completed

12. As the machine is turning, quickly and continuously raise and lower the grinder on and off the seat.   ☐ Task completed

13. Continue to grind until the seat is clean and free of defects.   ☐ Task completed

14. Check the fit of the valve with machinist's dye. Correct the fit as necessary. Describe what you need to do to have the valve properly fit the seat.

_____

_____

_____

_____

Instructor's Comments

_____

_____

ENGINE REPAIR JOB SHEET 17

Checking the Condition and Concentricity of a Valve Seat

Name _____ Station _____ Date _____

NATEF Correlation

This Job Sheet addresses the following NATEF task:

B.7. Check valve face-to-seat contact and valve seat concentricity (runout); determine necessary action.

Objective

Upon completion of this job sheet, you will be able to check the valve face-to-seat contact and valve seat concentricity (runout).

Tools and Materials

Machinist's rule or valve seat width scale

Service manual

Valve seat concentricity (runout) gauge

Selected valve guide arbors

Prussian blue machinist's dye

Protective Clothing

Goggles or safety glasses with side shields

Describe the vehicle being worked on:

Year _____ Make _____ Model _____

VIN _____ Engine type and size _____

PROCEDURE

1. Clean the valve seats in the cylinder head.   ☐ Task completed

2. Check the seats for cracks, excessive wear, and damage. Record your findings:

   _____

   _____

3. Check the seats for looseness in their bores. Record your findings:

   _____

   _____

4. Look up the specs for valve seat width in the service manual. Record the width for the exhaust and intake valve seats.

_____

_____

5. Summarize their condition based on your visual inspection.

_____

_____

6. With the machinist's rule or valve seat width scale, measure the width of the seats. Your measurements for each valve seat are:

_____

_____

7. Compare your measurements with the specs and summarize your conclusions and recommendations:

_____

_____

8. Select the correct size arbor and insert it into the valve guide.   ☐ Task completed

9. Place the concentricity gauge over the arbor.   ☐ Task completed

10. Slowly rotate the gauge while observing the readings on the gauge. What was the total runout shown on the gauge?

_____

_____

11. Compare your measurement to specs and give your conclusions.

_____

_____

12. Repeat the runout check on all valve seats and summarize your findings.

_____

_____

13. If runout was satisfactory, coat the appropriate valve face with Prussian blue dye.   ☐ Task completed

14. Place the valve into the appropriate seat and rotate it several times on the seat.   ☐ Task completed

15. Remove the valve from the seat and observe the marks from the dye.   ☐ Task completed

16. Is the contact area even? Is it in the correct location on the seat? Is the contact area the correct width?

_____

_____

17. Repeat this procedure on all of the seats.   ☐ Task completed

18. Summarize your findings and recommendations for service.

_____

_____

_____

_____

Instructor's Comments

_____

_____

_____

ENGINE REPAIR JOB SHEET 18

Check Valve Spring Assembled Height and Valve Stem Height

Name _____ Station _____ Date _____

NATEF Correlation

This Job Sheet addresses the following NATEF task:

B.8. Check valve spring assembled height and valve stem height; determine necessary action.

Objective

Upon completion of this job sheet, you will be able to measure the assembled height of a valve spring and the height of a valve stem.

Tools and Materials

Divider

Machinist rule

Valve spring tension tester

Service manual

Protective Clothing

Goggles or safety glasses with side shields

Describe the vehicle being worked on:

Year _____ Make _____ Model _____

VIN _____ Engine type and size _____

PROCEDURE

1. Prior to installing the valve and fitting valve springs, all other headwork   ☐ Task completed
   should be completed.

2. Install the valve into its proper valve guide.   ☐ Task completed

3. Install the valve retainer and keepers. Without the spring, these must be   ☐ Task completed
   held in place by hand.

4. While pulling up on the retainer, measure the distance between the bot-   ☐ Task completed
   tom of the retainer and the spring pad on the cylinder head with a
   divider.

5. Use a scale to determine the measurement expressed by the divider. The
   measurement was:

   _____

6. Compare this measurement with the specifications given in the service manual for installed spring height. What are the specifications for valve stem height?

   _____

   _____

7. If the measured installed height is greater than the specifications, a valve shim must be placed under the spring to correct the difference. Did you find a difference? What should you do? Show the math.

   _____

   _____

   _____

8. Spring tension must be checked at the installed spring height; therefore, if a shim is to be used, insert it under the spring on the valve spring tension gauge.   ☐ Task completed

9. Compress the spring into the installed height by pressing down on the tester's lever.   ☐ Task completed

10. The tension gauge will reflect the pressure of the spring when compressed to the installed or valve closed height. What were your readings?

    _____

    _____

11. Compare your measurement to the specifications. What are the specifications? How do your measurements differ from the specifications?

    _____

    _____

12. Now compress the spring to the open height specification. Use the rule on the gauge or a scale to measure the compressed height. What is the compressed height and what tension did you measure at that height?

    _____

    _____

13. Compare this reading to the specifications. Any pressure outside the pressure range given in the specifications indicates that the spring should be replaced. What are the specifications? What should you do with the valve spring?

    _____

    _____

14. After the tension and height have been checked, the spring can be installed on the valve stem.   ☐ Task completed

Instructor's Comments

_____

_____

_____

ENGINE REPAIR JOB SHEET 20

Adjust Valves on an OHC Engine

Name _____ Station _____ Date _____

NATEF Correlation

This Job Sheet addresses the following NATEF task:

B.11. Adjust valves (mechanical or hydraulic lifters).

Objective

Upon completion of this job sheet, you will be able to adjust the valves on an OHC engine.

Tools and Materials

Hand tools

Feeler gauge

Remote starter button

Allen wrench set

Service manual

Protective Clothing

Goggles or safety glasses with side shields

Describe the vehicle being worked on:

Year _____ Make _____ Model _____

VIN _____ Engine type and size _____

Describe what is used to maintain valve lash on the engine:

_____

PROCEDURE

1. Check the service manual to determine if the valves are supposed to be adjusted with a cold engine or when the engine is at normal operating temperature. ☐ Task completed

2. Remove necessary hoses and wires to allow the removal of the cam cover (valve cover). ☐ Task completed

3. Adjust the valves according to the procedure described in the service manual for the engine you are working on. ☐ Task completed

4. Re-install the cam cover, and connect any wires or hoses that were disconnected. ☐ Task completed

5. Start the engine to check the operation and check for oil leaks. ☐ Task completed

Problems Encountered

_____

_____

_____

Instructor's Comments

_____

_____

_____

ENGINE REPAIR JOB SHEET 21

Inspect and Replace Camshaft Drives

Name _____ Station _____ Date _____

NATEF Correlation

This Job Sheet addresses the following NATEF task:

B.12. Inspect camshaft drives (including gear wear and backlash, sprocket and chain wear); determine necessary action.

Objective

Upon completion of this job sheet, you will be able to inspect the camshaft drives, including gear wear and backlash, sprocket and chain wear.

Tools and Materials

Hand tools

Service manual

Protective Clothing

Goggles or safety glasses with side shields

Describe the engine being worked on:

Year _____ Make _____ Model _____

VIN _____ Engine type and size _____

_____

PROCEDURE

1. Disconnect the negative battery cable. ☐ Task completed

2. Remove the covers protecting the camshaft drive gears. ☐ Task completed

3. Check for proper alignment of the timing marks on the timing gears. ☐ Task completed

4. Remove the timing gears according to the service manual. ☐ Task completed

5. Clean the timing gears. ☐ Task completed

6. Inspect the gears for wear or cracks, and replace necessary parts. ☐ Task completed

7. After replacing any worn or damaged gears and drives, set the timing marks according to the service manual. ☐ Task completed

8. Rotate the engine through two complete rotations of the crankshaft. Check to see if the timing marks on the gears are still in proper alignment. ☐ Task completed

9. Reinstall all covers removed for access.   □ Task completed

10. Reconnect the negative battery cable and start the engine.   □ Task completed

11. Set the ignition timing to the manufacturer's specifications.   □ Task completed

Problems Encountered

_____

_____

_____

Instructor's Comments

_____

_____

_____

ENGINE REPAIR JOB SHEET 22

Replace a Timing Belt on an OHC Engine

Name _____ Station _____ Date _____

NATEF Correlation

This Job Sheet addresses the following NATEF task:

B.13. Inspect and replace timing belts(s), overhead camdrive sprockets, and tensioners; check belt/chain tension; adjust as necessary.

Objective

Upon completion of this job sheet, you will be able to inspect and replace timing belt(s), overhead cam-drive sprockets, and tensioners, as well as check belt tensions.

Tools and Materials

Basic hand tools

Paint stick or chalk

Belt tension gauge

Protective Clothing

Goggles or safety glasses with side shields

Describe the vehicle being worked on:

Year _____ Make _____ Model _____

VIN _____ Engine type and size _____

PROCEDURE

1. Disconnect the negative cable from the battery prior to beginning to remove and replace the timing belt. ☐ Task completed

2. Carefully remove the timing cover. Be careful not to distort or damage it while pulling it up. With the cover removed, check the immediate area around the belt for wires and other obstacles. If some are found, move them out of the way. What needed to be removed? Did you need to remove other drive belts?

   _____

   _____

   _____

3. Align the timing marks on the camshaft's sprocket with the mark on the cylinder head. If the marks are not obvious, use a paint stick or chalk to clearly mark them. ☐ Task completed

4. Carefully remove the crankshaft timing sensor and probe holder.   ☐ Task completed

5. Loosen the adjustment bolt on the belt tensioner pulley. It is normally not   ☐ Task completed
necessary to remove the tensioner assembly.

6. Slide the belt off the crankshaft sprocket. Be careful not to allow the crank-   ☐ Task completed
shaft pulley to rotate while doing this.

7. To remove the belt from the engine, the crankshaft pulley may need to be
removed to slip it off the crankshaft sprocket. Did you need to remove
the pulley? What did you need to do to remove the pulley?

_____

_____

_____

8. After the belt has been removed, inspect it for cracks and other damage.
Cracks will become more obvious if the belt is twisted slightly. Describe
any defects in the belt. NOTE: Timing belts are always replaced once
they have been removed. When may this not be true?

_____

_____

_____

9. To begin reassembly, place the belt around the crankshaft sprocket. Then   ☐ Task completed
reinstall the crankshaft pulley.

10. Make sure the timing marks on the crankshaft pulley are lined up with the   ☐ Task completed
marks on the engine block. If they are not, carefully rock the crankshaft
until the marks are lined up.

11. With the timing belt fitted onto the crankshaft sprocket and the crankshaft   ☐ Task completed
pulley tightened in place, the crankshaft timing sensor and probe can be
reinstalled.

12. Align the camshaft sprocket with the timing marks on the cylinder head.   ☐ Task completed
Then wrap the timing belt around the camshaft sprocket and allow the belt
tensioner to put a slight amount of pressure on the belt.

13. Adjust the tension as described in the service manual. Then rotate the
engine two complete turns. Recheck the tension. What are the specifications for belt tension? Why do you need to rotate the engine twice before
rechecking the tension?

_____

_____

_____

_____

14. Rotate the engine through two complete turns again, then check the align-   ☐ Task completed
ment marks on the camshaft and the crankshaft. Any deviation needs to
be corrected before the timing cover is reinstalled.

Instructor's Comments

ENGINE REPAIR JOB SHEET 24

Install Camshaft Bearings

Name _____ Station _____ Date _____

NATEF Correlation

This Job Sheet addresses the following NATEF task:

B.15. Inspect camshaft bearing surface for wear, damage, out-of-round, and alignment; determine necessary action.

Objective

Upon completion of this job sheet, you will be able to inspect and measure camshaft bearings for wear, damage, out-of-round, and alignment.

Tools and Materials

Bearings	Mandrel
Heavy ball peen hammer	Service manual
Light engine oil	Short or long driving bar

Protective Clothing

Goggles or safety glasses with side shields

Describe the engine being worked on:

Year _____ Make _____ Model _____

VIN _____ Engine type and size _____

PROCEDURE

1. Wipe the camshaft bores. Sort and lay out the new bearings in the correct order, from front to rear. Select either a short or long driving bar. Then select a mandrel closest to the cam bearing in size. ☐ Task completed

2. Place the bearing on the mandrel (it will fit loosely), and turn the bearing driver clockwise until the mandrel is snug in the bearing. Lubricate the bearing with light engine oil. ☐ Task completed

3. Start the bearing into the block by hand pressure. Make sure the oil passages align before driving the bearing into the block housing, and be sure to back off on the driving bar one-eighth turn. This will allow the 0.003- to 0.006-inch press fit. ☐ Task completed

4. Refer to the engine manufacturer's specifications for proper bearing positioning. Generally, the bearing will be seated correctly when the mandrel is flush with the face of the block. Some cam bearings must be installed further in the bore to align with the oil holes in the block. ☐ Task completed

5. Drive the bearing into the block by using sharp blows with a heavy ball peen hammer. Make sure the oil holes in the bearing and block are aligned. ☐ Task completed

   WARNING: *Always wear eye protection whenever striking a driver bar with a hammer. Keep the driver bar dressed to prevent chips from flying off.*

   WARNING: *Make sure the hammer handle is properly wedged to the hammer head. Do not use a hammer with a loose handle.*

6. Remove the mandrel by withdrawing the driving bar. Check for any nicks or burrs on the bearing and remove, if present. Follow the same procedure for the remainder of the bearings. ☐ Task completed

Problems Encountered

_____

_____

_____

Instructor's Comments

_____

_____

_____

ENGINE REPAIR JOB SHEET 25

Valve Timing Check

Name _____ Station _____ Date _____

NATEF Correlation

This Job Sheet addresses the following NATEF task:

B.16. Establish camshaft(s) timing and cam sensor indexing according to manufacturer's specifications and procedures.

Objective

Upon completion of this job sheet, you will be able to verify camshaft timing according to the manufacturer's specifications and procedures.

Tools and Materials

Spark plug socket

Compression gauge

Remote starter switch

Flashlight

Misc. hand tools

Large breaker bar

New rocker arm or camshaft cover gasket

Protective Clothing

Goggles or safety glasses with side shields

Describe the vehicle being worked on:

Year _____ Make _____ Model _____

VIN _____ Engine type and size _____

PROCEDURE

1. If the timing belt or chain has slipped on the camshaft sprocket, the engine may fail to start because the valves are not properly timed in relation to the crankshaft. When the timing belt or chain has only slipped a few cogs on the camshaft sprocket, the engine has a lack of power, and fuel consumption is excessive. To check valve timing, begin by removing the spark plug from number 1 cylinder. ☐ Task completed

2. Disconnect the positive primary wire from the ignition coil to disable the ignition system. ☐ Task completed

3. Connect a remote control switch to the starter solenoid terminal and the battery terminal on the solenoid. ☐ Task completed

4. Place your thumb on top of the spark plug hole at cylinder #1. If this hole is not accessible, place a compression gauge in the opening.   ☐ Task completed

5. Crank the engine until compression is felt at the spark plug hole. Then, slowly crank the engine until the timing mark lines up with the zero-degree position on the timing indicator. The number 1 piston is now at TDC on the compression stroke. On many engines, the timing mark is on the crankshaft pulley, and the timing indicator is mounted above the pulley.   ☐ Task completed

6. Now, slowly crank the engine for one revolution until the timing mark lines up with the zero-degree position on the timing indicator. The number 1 piston is now at TDC on the exhaust stroke.   ☐ Task completed

7. Remove the rocker arm or camshaft cover and install a breaker bar and socket on the crankshaft pulley nut. Observe the valve action while rotating the crankshaft about 30 degrees before and after TDC on the exhaust stroke. In this crankshaft position, the exhaust valve should close a few degrees after TDC on the exhaust stroke, and the intake valve should open a few degrees before TDC on the exhaust stroke. Is this what you observed?   ☐ Task completed

   _____

8. If the valves did not open properly in relation to the crankshaft position, the valve timing is not correct. What should you do to correct it?

   _____

   _____

   _____

   _____

9. If the timing was correct, reinstall the rocker arm or camshaft cover with a new gasket. Tighten the attaching bolts to the proper specification. The recommended torque is _____

10. Reinstall the spark plug and tighten it to the proper specification. The recommended torque is _____

Problems Encountered

_____

_____

_____

Instructor's Comments

_____

_____

_____

ENGINE REPAIR JOB SHEET 26

Service and Install Oil Pump and Oil Pan

Name _____ Station _____ Date _____

NATEF Correlation

This Job Sheet addresses the following NATEF task:

> **D.2.** Inspect oil pump gears or rotors, housing, pressure relief devices, and pump drive; perform necessary action.

This Job Sheet also relates to the following NATEF task:

> **C.1.** Disassemble engine block; clean and prepare components for inspection and reassembly.

Objective

Upon completion of this job sheet, you will be able to inspect and replace the pans, covers, gaskets, and seals as well as inspect the oil pump gears or rotors, housing, pressure relief devices, and pump drive.

Tools and Materials

Feeler gauge   Service manual

Measuring scale   Sockets and ratchet

Outside micrometer   Straightedge

Pan gaskets   Torque wrench

Protective Clothing

Goggles or safety glasses with side shields

Describe the engine being worked on:

Year _____ Make _____ Model _____

VIN _____ Engine type and size _____

Describe the type of oil pump:

_____

PROCEDURE

1. Using a service manual, locate the required specifications for the Report Sheet for Oil Pump Service and enter them there. ☐ Task completed

   WARNING: *Always wear safety goggles or glasses with side shields when doing this task. The oil pressure relief valve is controlled by spring pressure, and the retainer may fly out of the housing, causing severe injury.*

2. Disassemble the oil pump and relief valve assembly in clean solvent and allow them to dry. ☐ Task completed

3. Place a straightedge across the pump cover, and try to push a 0.002-in. feeler gauge between the cover and straightedge. If the gauge fits, the cover is worn. Record your results on the Report Sheet for Oil Pump Service. □ Task completed

4. Measure the thickness and diameter of the rotors or gears. Compare the measurements with specifications. Measurements smaller than the specifications mean the pump must be replaced. Record your results on the Report Sheet for Oil Pump Service. □ Task completed

5. Assemble the gears or rotors into the pump body. Measure the clearance between them with a feeler gauge. Record your results on the Report Sheet for Oil Pump Service. If the clearance is larger than specifications, replace the pump. □ Task completed

6. Position a straightedge across the gears or rotors, and measure the clearance with a feeler gauge. If the measurement is larger than specifications, replace the pump. Record your results on the Report Sheet for Oil Pump Service. □ Task completed

7. Measure the relief valve spring height and record your findings on the Report Sheet for Oil Pump Service. □ Task completed

8. Lubricate all pump parts and reassemble pump. Tighten all bolts to specifications. □ Task completed

9. Install the pump drive or extension and install the pump on the engine. Torque the mounting bolts to specifications. □ Task completed

10. Inspect the sealing surface of the pan and cylinder block. Install new gaskets and seals on the oil pan. Use gasket sealer or sealant, as specified. Install the pan and tighten the pan bolts to specifications. □ Task completed

11. Install a new pan drain-plug gasket and install the drain plug. Tighten to specifications. □ Task completed

Problems Encountered

_____

_____

_____

Instructor's Comments

_____

_____

_____

REPORT SHEET FOR OIL PUMP SERVICE

Measuring	Specifications	Actual
Cover warpage		
Rotor diameter		
Rotor thickness		
Rotor clearance (backlash)		
Gear (rotor)-to-cover clearance		
Relief spring free length		
	Torque Specifications	
Oil pump cover bolts		
Oil pump mounting bolts		
Oil pan bolts		
Oil pan drain plug		

Conclusions and Recommendations _____

_____

_____

ENGINE REPAIR JOB SHEET 27

Engine Block Inspection

Name _____ Station _____ Date _____

NATEF Correlation

This Job Sheet addresses the following NATEF task:

> **C.2.** Inspect engine block for visible cracks, passage condition, core and gallery plug condition, and surface warpage; determine necessary action.

Objective

Upon completion of this job sheet, you will be able to inspect an engine block for visible cracks, check passage condition, check core and gallery plug condition, and surface warpage.

Tools and Materials

Tap set

Straightedge

Feeler gauge set

Miscellaneous wire brushes

Protective Clothing

Goggles or safety glasses with side shields

Describe the vehicle being worked on:

Year _____ Make _____ Model _____

VIN _____ Engine type and size _____

PROCEDURE

1. Place the block in a position in which you can easily clean the threaded □ Task completed
   bores. This may involve rotating the block periodically while doing this.

2. Run the correct-size tap through all threaded bores. Clean the tap after
   each use. Did you have a hard time running the tap through any bores?
   What was the problem?

   _____

   _____

3. Check to make sure the threads are in good condition. If they are not,
   they may need to be replaced. Record your findings and conclusions:

   _____

   _____

   _____

4. Carefully look all around the block for evidence of coolant and/or oil leaks. Summarize your results:

_____

_____

_____

5. Check the seal around each core and gallery plug. Are there signs of leakage? If so, what must be done?

_____

_____

_____

6. Remove the core and gallery plugs. What types of plugs are used in this engine block? What did you need to do to remove the plugs?

_____

_____

_____

7. After the plugs have been removed, inspect the sealing surface for the plugs and record their condition:

_____

_____

8. Run a properly sized wire brush through the water and oil passages.   ☐ Task completed
   Make sure all debris is removed.

9. Carefully inspect the areas around all bores, threaded or non-threaded, for evidence of cracks. Summarize your results:

_____

_____

_____

_____

10. Place the straightedge diagonally across the deck surface.   ☐ Task completed

11. The amount of warpage is determined by the size of feeler gauge you can fit into the gap between the straightedge and the deck. What was the largest feeler gauge blade you could insert?

_____

12. Move the straightedge to the other diagonal on the deck surface and repeat the above procedure. What was the largest feeler gauge blade you could insert?

_____

13. How much warpage is there on the deck surface? What do you recommend?

_____

_____

_____

14. After the block has been thoroughly inspected and cleaned, it is ready for   ☐ Task completed
further service and then assembly.

Instructor's Comments

_____

_____

_____

ENGINE REPAIR JOB SHEET 28

How to Repair and Replace Damaged Threads

Name _____ Station _____ Date _____

NATEF Correlation

This Job Sheet addresses the following NATEF task:

C.3. Inspect internal and external threads; restore as needed (includes installing thread inserts).

Objective

Upon completion of this job sheet, you will be able to inspect internal and external threads, as well as install thread inserts.

Tools and Materials

Tap and die set   Drill

Heli-coil set   Drill bits

Center punch

Protective Clothing

Face shield

Goggles or safety glasses with side shields

PROCEDURE TO REPAIR DAMAGED THREADS

1. Using a thread gauge, determine the proper thread pattern of the damaged threads.   ☐ Task completed

2. Thoroughly clean the threads to be repaired.   ☐ Task completed

3. Select the proper size tap to repair internal threads or die to repair external threads.   ☐ Task completed

4. Align the repair tool with the damaged threads, and slowly rotate the tool in the direction of the threads. This tool will clean and straighten the damaged threads.   ☐ Task completed

5. When you have reached the end of the damaged threads, remove the tool by reversing its rotation.   ☐ Task completed

   CAUTION: *Use a light oil to enable the taps and dies to cut easily and cleanly.*

PROCEDURE TO REPLACE DAMAGED THREADS

When internal threads in an object are damaged and cannot be repaired, they are sometimes replaced with a heli-coil.

1. Select the proper size heli-coil kit and use the drill bit size recommended to drill out the hole. ☐ Task completed

2. Use the tap furnished to thread the oversized hole. ☐ Task completed

3. Use the provided tool to install the new thread kit. ☐ Task completed

Problems Encountered

_____

_____

_____

Instructor's Comments

_____

_____

_____

ENGINE REPAIR JOB SHEET 29

Remove Cylinder Ring Ridge

Name _____ Station _____ Date _____

NATEF Correlation

This Job Sheet is related to the following NATEF task:

C.4. Inspect and measure cylinder walls for damage and wear; determine necessary action.

Objective

Upon completion of this job sheet, you will be able to remove the cylinder wall ridges with a ridge reamer.

Tools and Materials

Oily rag

Ridge removal tool

Wrench

Protective Clothing

Safety goggles or glasses with side shields

Describe the vehicle being worked on:

Year _____ Make _____ Model _____

VIN _____ Engine type and size _____

PROCEDURE

1. Rotate each piston to bottom dead center and inspect the cylinder to see if there is a ring ridge. Use your fingernail to check for a ridge. If it catches on the cylinder wall, the ridge must be removed. ☐ Task completed

2. Select the correct-size ridge reamer and position it in the cylinder with the piston at bottom dead center. Adjust the cutter against the cylinder walls according to the instructions furnished with the tool. ☐ Task completed

3. Use a wrench to turn the tool in a clockwise direction. ☐ Task completed

   WARNING: *Always wear safety goggles or glasses with side shields when completing this task.*

   Rotate the tool around the cylinder until the ridge is removed. Be careful not to remove more than the ridge.

4. Repeat this operation in each of the other cylinders. ☐ Task completed

5. Use an oily rag to remove the cuttings from each cylinder. ☐ Task completed

Problems Encountered

_____

_____

_____

Instructor's Comments

_____

_____

_____

ENGINE REPAIR JOB SHEET 30

Measure Cylinder Bore

Name _____ Station _____ Date _____

NATEF Correlation

This Job Sheet addresses the following NATEF task:

C.4. Inspect and measure cylinder walls for damage and wear; determine necessary action.

Objective

Upon completion of this job sheet, you will be able to inspect and measure cylinder walls for damage and wear.

Tools and Materials

Outside micrometer

Service manual

Telescoping gauge

Protective Clothing

Safety goggles or glasses with side shields

Describe the engine being worked on:

Year _____ Make _____ Model _____

VIN _____ Engine type and size _____

PROCEDURE

1. Using the appropriate service manual, look up the specifications for standard bore size, tolerances for out-of-roundness, and taper. Record these specifications on the Report Sheet for Measuring Cylinder Bore.   ☐ Task completed

2. Release the lock screw at the end of the handle of the telescoping gauge. Compress the plungers and, using the lock screw, secure them in the retracted position. Place the telescoping gauge into the bore, in an area below ring travel and 90 degrees to the crankshaft.   ☐ Task completed

3. Release the plungers by loosening the lock screw. Allow the plungers to expand until they contact the bore walls. Rock the telescoping gauge back and forth and side to side to check for the correct resistance. Lock the plungers into position using the lock screw. Carefully retract the telescoping gauge from the bore.   ☐ Task completed

4. Use an outside micrometer to measure the distance between the two plunger faces. Record the reading obtained for each cylinder on the Report Sheet for Measuring Cylinder Bore.   ☐ Task completed

5. Measure the diameter of the cylinder at the highest point of piston ring travel. The difference between this measurement and the diameter below ring travel is the amount the cylinder is tapered. Record the amount of taper for each cylinder on the Report Sheet for Measuring Cylinder Bore.   ☐ Task completed

6. Measure for cylinder out-of-roundness by first measuring the bore at a point parallel to the piston pin, then at right angles to the piston. The difference between these measurements is the amount of out-of-roundness. Measure for out-of-roundness at three locations in each bore (top, middle, and bottom). Record the worst condition for each cylinder on the Report Sheet for Measuring Cylinder Bore.   ☐ Task completed

7. Perform a visual inspection of the block and record your findings on the Report Sheet for Measuring Cylinder Bore.   ☐ Task completed

Problems Encountered

_____

_____

_____

Instructor's Comments

_____

_____

_____

Name _____   Station _____   Date _____

REPORT SHEET FOR MEASURING CYLINDER BORE								
Cylinder No.	*1*	*2*	*3*	*4*	*5*	*6*	*7*	*8*
Standard bore size								
Actual bore size								
Taper limits								
Actual taper								
Out-of-round limits								
Actual out-of-roundness								

Visual Inspection

Cracks and scores	☐ OK	☐ Not serviceable
Core-hole plugs	☐ OK	☐ Not serviceable
Lifter bores	☐ OK	☐ Not serviceable
Water jackets	☐ OK	☐ Not serviceable
Machined surfaces	☐ OK	☐ Not serviceable
Bolt holes and threads	☐ OK	☐ Not serviceable
Oil galleries	☐ OK	☐ Not serviceable
Oil passages	☐ OK	☐ Not serviceable

Conclusions and Recommendations _____

_____

_____

ENGINE REPAIR JOB SHEET 31

Deglaze and Clean Cylinder Walls

Name _____ Station _____ Date _____

NATEF Correlation

This Job Sheet addresses the following NATEF task:

C.5. Deglaze and clean cylinder walls.

Objective

Upon completion of this job sheet, you will be able to deglaze and clean the cylinder walls in an engine block.

Tools and Materials

Glaze breaker

Variable-speed electric drill or honing machine

Large, round stiff-bristled brush

Clean lint-free cloth

Protective Clothing

Goggles or safety glasses with side shields

Describe the vehicle being worked on:

Year _____ Make _____ Model _____

VIN _____ Engine type and size _____

PROCEDURE

1. Carefully inspect and measure the cylinder bores for surface condition, taper, and out-of-roundness. Record your findings here:

   _____

   _____

   _____

2. If the bores are within acceptable limits, the cylinder walls only need to be deglazed. What causes glaze on the cylinder walls?

   _____

   _____

3. State what you are going to use to spin the glaze breaker:

_____

_____

4. What grit of glaze breaker are you going to use?

_____

5. Run the glaze breaker up and down the bores.   ☐ Task completed

6. After deglazing, use plenty of hot, soapy water and a stiff-bristle brush to   ☐ Task completed
   clean the bores.

7. Wipe the walls dry with a lint-free cloth.   ☐ Task completed

8. Rinse the block with water and dry it again.   ☐ Task completed

9. Lightly coat the cylinder walls with clean, light engine oil. Why should
   you do this?

_____

_____

Instructor's Comments

_____

_____

_____

ENGINE REPAIR JOB SHEET 32

Inspect Camshaft Bearings

Name _____ Station _____ Date _____

NATEF Correlation

This Job Sheet addresses the following NATEF task:

C.6. Inspect and measure camshaft bearings for wear, damage, out-of-round, and alignment; determine necessary action.

Objective

Upon completion of this job sheet, you will be able to inspect and measure the camshaft bearings for wear, damage, out-of-roundness, and alignment.

Tools and Materials

Telescoping gauge

Straight edge

Micrometer

Feeler gauge

Service manual

Protective Clothing

Goggles or safety glasses with side shields

Describe the vehicle being worked on:

Year _____ Make _____ Model _____

VIN _____ Engine type and size _____

PROCEDURE

1. Carefully inspect the camshaft bearings. Are they in the block or in the cylinder head? Describe their condition.

   _____

   _____

2. Look up the spec for the ID of the camshaft bearings and record it here:

   _____

   _____

3. With a telescoping gauge and micrometer, measure the ID of each bearing and summarize your findings here:

   _____

   _____

4. Compare your measurements to the specs and state your conclusions.

_____

_____

5. Check the ID of the bearings at 90 degree intervals. Are the ID measurements the same or are there signs of out-of-roundness? Summarize your findings.

_____

_____

6. If the bearings were found not to be out-of round, lay a straightedge across the bottom of the bearings. Check to see if the bores are aligned by attempting to put a feeler gauge under the straightedge in each of the bores. Summarize your findings here:

_____

_____

7. What service do you recommend after having done these checks?

_____

_____

Instructor's Comments

_____

_____

_____

ENGINE REPAIR JOB SHEET 33

Measure Crankshaft Journals

Name _____ Station _____ Date _____

NATEF Correlation

This Job Sheet addresses the following NATEF task:

C.7. Inspect crankshaft for end play, straightness, journal damage, keyway damage, thrust flange and sealing surface condition, and visual surface cracks; check oil passage condition; measure journal wear; check crankshaft sensor reluctor ring (where applicable).

Objective

Upon completion of this job sheet, you will be able to inspect the crankshaft for surface cracks and journal damage, check the oil passage condition, and measure for journal wear.

Tools and Materials

Outside micrometer

Service manual

Protective Clothing

Safety goggles or glasses with side shields

Describe the engine being worked on:

Year _____ Make _____ Model _____

VIN _____ Engine type and size _____

PROCEDURE

1. Using a service manual, look up the specifications for standard crankshaft size and tolerances for normal wear. Record these specifications on the Report Sheet for Measuring Crankshaft Journals. ☐ Task completed

2. Check the crankshaft and use proper procedures to clean it before beginning the measurement process. ☐ Task completed

3. Using the proper size outside micrometer, check the number 1 main bearing journal twice at each end of the journal, once horizontal to the crankshaft and once vertical. Record all of the measurements on the report sheet. (**Note:** If these measurements are different, the crankshaft main bearing journal is out-of-round or tapered and the crankshaft should be machined.) ☐ Task completed

4. Repeat step 3 for each of the remaining main bearing journals. ☐ Task completed

5. Measure the number 1 connecting rod journal twice at each end of the journal, once horizontal to the crankshaft and once vertical. Record all of the measurements on the report sheet. (**Note:** If these measurements are different, the crankshaft connecting rod journal is out-of-round or tapered, and the crankshaft should be machined.)   ☐ Task completed

6. Repeat step 5 for each of the remaining connecting rod journals.   ☐ Task completed

7. Compare the measurements of the crankshaft journals with the specifications. If the measurements are within specifications, the crankshaft can be reinstalled in the engine. If the measurements are not within factory specifications, the crankshaft must be machined before it is reinstalled in the engine.   ☐ Task completed

Problems Encountered

_____

_____

_____

Instructor's Comments

_____

_____

_____

Name _____   Station _____   Date _____

REPORT SHEET FOR MEASURING CRANKSHAFT MEASUREMENTS							
Main Journal Number	1	2	3	4	5	6	7
Standard journal size							
Actual journal size							
Taper limits							
Actual taper							
Out-of-round limits							
Actual out-of-roundness							
Connecting Rod Journal Number	1	2	3	4	5	6	
Standard journal size							
Actual journal size							
Taper limits							
Actual taper							
Out-of-round limits							
Actual out-of-roundness							

Conclusions and Recommendations _____

_____

_____

ENGINE REPAIR JOB SHEET 34

Check Crankshaft End Play

Name _____ Station _____ Date _____

NATEF Correlation

This Job Sheet addresses the following NATEF task:

C.8. Inspect and measure main and connecting rod bearings for damage, clearance, and end play; determine necessary action (includes the proper selection of bearing)

Objective

Upon completion of this job sheet, you will be able to inspect and measure the main bearings and connecting rod bearings for damage, clearance, and end play.

Tools and Materials

Bracket   Light engine oil

Dial indicator   Service manual

Feeler gauge set   Two pry bars

Protective Clothing

Goggles or safety glasses with side shields

Steel-toed shoes

Describe the engine being worked on:

Year _____ Make _____ Model _____

VIN _____ Engine type and size _____

PROCEDURE

1. Following the manufacturer's recommended procedures presented in a service manual, install the rear main oil seal in the block and bearing cap. ☐ Task completed

2. Thoroughly lubricate all bearings and rear main oil seal with engine oil. Install the main bearing halves into the main bearing bores in the block. Make sure the oil holes in the bearing halves line up with the oil holes in the block. ☐ Task completed

3. Carefully install the crankshaft into the block. ☐ Task completed

4. Install the main bearing caps with bearings, being careful to match the location numbers on the cap with the block. ☐ Task completed

5. Tighten the main bearing cap bolts one at a time in three steps until the full, specified torque is obtained. Do not torque the thrust bearing cap at this time. ☐ Task completed

6. Pry crankshaft back and forth to align the thrust surfaces at the thrust bearing. While prying the crankshaft forward, pry the thrust bearing cap rearward. When assured that the surfaces are aligned, torque the cap bolts to specifications in three steps. ☐ Task completed

7. Using a service manual, locate the minimum and maximum crankshaft end play specifications. Record specifications. ☐ Task completed

   Minimum specification _____

   Maximum specification _____

8. Pry the crankshaft toward the front of the engine. Install a dial indicator and bracket so that its plunger rests against the crankshaft flange and the indicator axis is parallel to the crankshaft axis. ☐ Task completed

   NOTE: *This procedure can be accomplished using a feeler gauge if a dial indicator is not available. This check would be made between the thrust bearing and the thrust surfaces on the crankshaft.*

9. Set the dial indicator at zero. Pry the crankshaft rearward. Note the reading on the dial indicator. ☐ Task completed

10. Record your dial indicator reading: ☐ Task completed

    _____

11. If your results are not within manufacturer's specifications, list your conclusions and recommendations. ☐ Task completed

Problems Encountered

_____

_____

_____

Instructor's Comments

_____

_____

_____

ENGINE REPAIR JOB SHEET 35

Inspecting Connecting Rod Bearings

Name _____ Station _____ Date _____

NATEF Correlation

This Job Sheet addresses the following NATEF task:

C.9. Identify piston and bearing wear patterns that indicate connecting rod alignment and main bearing bore problems; inspect rod alignment and bearing bore condition.

Objective

Upon completion of this job sheet, you will be able to identify piston and bearing wear patterns that indicate connecting rod misalignment and main bearing bore problems.

Tools and Materials

Clean rags

Telescoping gauge set

Micrometer

Protective Clothing

Goggles or safety glasses with side shields

Describe the vehicle being worked on:

Year _____ Make _____ Model _____

VIN _____ Engine type and size _____

PROCEDURE

1. Wipe the bores of the connecting rods down with a clean cloth, then carefully inspect the big end bearings in each connecting rod and summarize your findings:

   _____

   _____

2. Carefully inspect the small end bearings in each connecting rod and summarize your findings:

   _____

   _____

3. Did either end of any bearing show signs of wear on one side only? What would this indicate?

   _____

   _____

4. With the telescoping gauge and a micrometer, measure the bore of the big end of the rod with the bearing in place. Begin by measuring the ID at both sides and in the center of the rod. Do this to each rod and summarize your findings.

_____

_____

5. With the telescoping gauge and a micrometer, measure the bore of the small end of the rod with the bearing in place. Begin by measuring the ID at both sides and in the center of the rod. Do this to each rod and summarize your findings.

_____

_____

6. Check the ID of the big end of each rod with the bearing in place. Place the telescoping gauge in the center of the bore and then measure at 90 degree intervals. Summarize your findings:

_____

_____

7. Check the ID of the small end of each rod with the bearing in place. Place the telescoping gauge in the center of the bore and then measure at 90-degree intervals. Summarize your findings:

_____

_____

8. What did the above checks look for?

_____

_____

9. Look up the specs for the correct ID of both connecting rod bores for this engine. Record them here:

_____

_____

10. Carefully inspect the bores of each connecting rod without the bearings and summarize your findings:

_____

_____

11. Based on all of these checks, what are your recommendations for service to the connecting rods?

_____

_____

_____

_____

Instructor's Comments

_____

_____

_____

ENGINE REPAIR JOB SHEET 36

Checking and Servicing Pistons and Pins

Name _____ Station _____ Date _____

NATEF Correlation

This Job Sheet addresses the following NATEF task:

C.10. Inspect and measure pistons; determine necessary action.

Objective

Upon completion of this job sheet, you will be able to inspect, measure, and service pistons and pins.

Tools and Materials

Scraper   Small hydraulic press

Cold tank   Variety of drivers and adapters

Ring groove cleaner   Feeler gauge

Protective Clothing

Goggles or safety glasses with side shields

Describe the vehicle being worked on:

Year _____ Make _____ Model _____

VIN _____ Engine type and size _____

PROCEDURE

1. Thoroughly remove the carbon from the top of the pistons. Be careful not to remove any metal from the piston. What did you need to do in order to get the carbon off?

   _____

   _____

2. Remove the piston rings. Clean and inspect the ring grooves. Make sure all of the openings at the rear of the grooves are free of carbon and/or debris.   ☐ Task completed

3. After the piston is cleaned, determine if the piston is offset on the rod. If so, what mark did the manufacturer use to indicate the direction of installation? If the piston is offset from the rod and there are no markings, make some. What could you use to do that?

   _____

   _____

4. Examine the piston pin and describe how it is retained. Describe that here:

_____

_____

5. Remove the piston from the rod according to the type of piston pin used.　□ Task completed

Full-Floating Pins

   a. Remove the snap rings from the pin boss. Note the direction the snap rings were installed in the boss. Describe that here:

   _____

   _____

   b. Push the pin out of the piston and connecting rod. Did you have any difficulty pushing the pin out?

   _____

   c. Carefully inspect the snap ring groove. Record your findings:

   _____

   _____

Press-Fit Pins

   a. Set the correct-size lower adapter onto the base of the press and place the piston and rod assembly on it. Does the adapter support the piston and will it allow enough room for the pin as it is pressed out?

   _____

   b. Choose the correct driver and set it into the pin. Does the driver fully contact the pin and will it clear the pin bore as the pin is pressed out?

   _____

   c. Press the pin out and separate the rod from the piston.　□ Task completed

   d. Examine the piston pin bore and describe the condition.

   _____

   _____

6. Visually inspect the piston head for any damage, then describe its condition.

_____

_____

7. Visually inspect the piston skirts for indications of scuffing. What did you find?

_____

_____

8. Measure the ring groove for wear by inserting a new piston ring backward in the groove and measure the clearance between the groove and the ring with a feeler gauge. The measurement was: _____

9. Compare this reading with the specifications. The specifications are: _____ What does that indicate?

   _____

   _____

10. Measure the bore for the piston pin and compare your reading to the specifications. Your measurement was: _____ The specifications call for a bore of: _____ Is there is a difference and, if so, what does that indicate?

    _____

    _____

11. Check the bore for taper, out-of-roundness, and parallelism. How did you do that?

    _____

    _____

12. Visually inspect the piston pin and record your observations.

    _____

    _____

13. Measure the diameter of the piston pin and compare your reading to the specifications. Your measurement was: _____ The specifications call for a bore of: _____ Is there is a difference and, if so, what does that indicate?

    _____

    _____

14. Inspect the small bore of the connecting rod and record the results of the inspection.

    _____

    _____

15. Measure the inside diameter of the small bore and compare your reading to the specifications. Your measurement was: _____ The specifications call for a bore of: _____ Is there is a difference, and, if so, what does that indicate?

    _____

    _____

16. Check the bore for taper, out-of-roundness, and parallelism. How did you do that?

_____

_____

17. Replace or recondition any part that needs to be. What parts need to be serviced prior to assembling the rod and piston?

_____

_____

18. Assemble the piston to the rod, according to the type of piston pin used.   ☐ Task completed

Full-Floating Pins

a. Coat the pin, piston pin bore, and small-end bore with light oil.   ☐ Task completed

b. Fit the piston pin into one side of the piston bore. Make sure any markings that indicate proper piston direction are aligned.   ☐ Task completed

c. Position the small-end of the rod in line with the piston pin bore.   ☐ Task completed

d. Push the pin through the small-end bore and into the other side of the piston's pin bore.   ☐ Task completed

e. Make sure the pin is free to rotate in both bores.   ☐ Task completed

f. Install new snap rings, making sure they face in the correct direction and are fully seated in their grooves.   ☐ Task completed

Press-Fit Pins

a. Make sure there is a slight chamfer on the outside of the small-end bore of the connecting rod.   ☐ Task completed

b. Coat the pin, piston pin bore, and small-end bore with light oil.   ☐ Task completed

c. Select the correct press adapter and set it on the base of the press. Does the adapter support the piston?

_____

d. Slide the piston pin into the top piston boss. Make sure any markings that indicate proper piston direction are aligned.   ☐ Task completed

e. Push the pin in until it makes contact and is centered in on the small-end bore of the rod.   ☐ Task completed

f. Choose the correct driver and set it into the pin.   ☐ Task completed

g. Tap the pin with a soft-faced mallet to start it into the connecting rod.   ☐ Task completed

h. Press the pin in until it makes contact with the stop.   ☐ Task completed

19. Describe any difficulties you had with this procedure:

_____

_____

Instructor's Comments

_____

_____

_____

ENGINE REPAIR JOB SHEET 37

Install Pistons, Connecting Rods, and Harmonic Balancer

Name _____ Station _____ Date _____

NATEF Correlation

This Job Sheet addresses the following NATEF tasks:

C.12. Inspect, measure, and install piston rings.

C.14. Inspect, repair or replace crankshaft vibration damper (harmonic balancer).

Objective

Upon completion of this job sheet, you will be able to inspect, measure and install piston rings as well as inspect, repair or replace the crankshaft vibration damper (harmonic balancer).

Tools and Materials

Allen wrench

Anaerobic thread-locking compound

Compressor tool

Feeler gauges

Flat-blade screwdriver

Light engine oil

Plastic mallet

Plastigage

Rubber or aluminum protectors or guides

Service manual

Protective Clothing

Goggles or safety glasses with side shields

Describe the engine being worked on:

Year _____ Make _____ Model _____

VIN _____ Engine type and size _____

PROCEDURE

1. Place rubber or aluminum protectors or guides over the threaded section of the rod bolts. Lightly coat the piston, rings, cylinder wall, crankpin, and compressor tool with light engine oil. Do not coat the rod bearings with oil at this time.   ☐ Task completed

2. Be sure that the ring gaps are located on the piston in the positions recommended in the service manual.   ☐ Task completed

3. Expand the compressor tool around the piston rings. Position the steps on the compressor tool downward. Tighten the compressor tool with an Allen wrench to compress the piston rings.   ☐ Task completed

4. Rotate the crankshaft until the crankpin is at its lowest level (BDC). Place the piston/rod assembly into the cylinder bore until the steps on the compressor tool contact the cylinder block deck. Make sure the piston reference mark is in the correct relation to the front of the engine.   ☐ Task completed

5. Remove the protective covering from the rod bolts. Lightly tap on the head of the piston with a mallet handle or block of wood until piston enters the cylinder bore.   ☐ Task completed

6. Push the piston down the bore while making sure the connecting rod fits into place on the crankpin.   ☐ Task completed

7. Look in the service manual for the specs on the oil clearance for connecting rod bearings and the torque specs for the rod cap bolts. What are these specs?

    _____

    _____

8. Check the rod bearing oil clearances on all rods with Plastigage. Record your results.

    _____

    _____

9. Based on the above, what do you recommend?

    _____

    _____

10. If the clearances are within specs, remove all cap bolts and push the piston and rod assembly into the appropriate cylinder bore. Lubricate the bearings and crankpins.   ☐ Task completed

11. Seat the connecting rod yoke onto the crankshaft.   ☐ Task completed

12. Position the matching connecting rod cap onto each rod and finger-tighten the rod nuts. When doing this, what do you need to check?

    _____

    _____

13. Gently tap each cap with the plastic mallet to seat it against the crankshaft and connecting rod.   ☐ Task completed

14. Torque the rod cap bolts to specs.   ☐ Task completed

15. Look in the service manual for the connecting rod side clearance specs. Record them here.

    _____

    _____

16. Measure the side clearance of each rod with a feeler gauge and record your findings.

    _____

    _____

17. Compare your measurements to specs and state your service recommendations.

_____

_____

18. If all clearances are correct, prepare to install the harmonic balancer to the crankshaft.   □ Task completed

19. Visually inspect the balancer for signs of wear on its center bore. Record your findings.

_____

_____

20. Inspect the rubber insulator for deterioration and twisting. Record your findings.

_____

_____

21. Based on these checks, does the balancer need to be replaced? Why or why not?

_____

_____

22. Carefully set the balancer over the snout of the crankshaft with the key and keyway aligned.   □ Task completed

23. Press the balancer onto the crankshaft, being careful not to move the key.   □ Task completed

24. Look up the torque spec for the balancer attaching bolt. The spec is:

_____

25. Tighten the balancer.   □ Task completed

Instructor's Comments

_____

_____

_____

ENGINE REPAIR JOB SHEET 38

Reassemble Engine

Name _____ Station _____ Date _____

NATEF Correlation

This Job Sheet addresses the following NATEF task:

C.15. Assemble the engine using gaskets, seals, and formed-in-place (tube-applied) sealants, thread sealers, etc. according to manufacturer's specifications.

Objective

Upon completion of this job sheet, you will be able to reassemble the engine and mount its components using the correct gaskets and sealants.

Tools and Materials

Straightedge   Hydraulic press

Flashlight   Drivers

Feeler gauge   Block of wood

Protective Clothing

Goggles or safety glasses with side shields

Describe the vehicle being worked on:

Year _____ Make _____ Model _____

VIN _____ Engine type and size _____

PROCEDURE

WARNING: *Make sure every sealant you use on today's engines is oxygen sensor safe.*

1. Visually inspect the bolts. Threads must be clean and undamaged. Discard all bolts that are not acceptable. Describe the condition of the bolts:

   _____

   _____

2. Gather new gaskets for the engine. Never reuse old gaskets. Even if the old gasket appears to be in good condition, it will never seal as well as a new one. Protect the new gaskets by keeping them in their packages until it is time to install them.   ☐ Task completed

3. Make sure all surfaces are free of dirt, oil deposits, rust, old sealer, and gasket material.   ☐ Task completed

4. Check the condition of the balancer shaft or crankshaft pulley hub. Make sure the surface is smooth. If the surface is not smooth, the seal will not be able to seal. What can be done if the pulley hub has a groove in it?

_____

_____

5. Before installing the oil pan and gasket, check the flanges for warpage. Use a straightedge or lay the pan, flange side down, on a flat surface with a flashlight underneath it to spot uneven edges. Carefully check the flange around bolt holes. What were the results of this check? If the flange is distorted, what should you do?

_____

_____

6. Once it has been determined that the flanges are flat, install the oil pan   ☐ Task completed
   with a new gasket.

7. When replacing the timing cover, remove the old gaskets and seals from   ☐ Task completed
   the timing cover and engine block.

8. Install a new crankshaft seal using a press, seal driver, or hammer and a   ☐ Task completed
   clean block of wood. When installing the seal, be sure to support the cover underneath to prevent damage.

9. If the timing cover extends over the front lip of the oil pan, the front portion of the oil pan gasket will be exposed. With a sharp knife or razor blade, carefully cut off the front exposed portion of the oil pan gasket. Did you need to do this?

_____

10. Apply a light coating of adhesive or sealant on the timing cover and position the gasket on the cover. Finally, mount the timing cover and torque the bolts to specifications. Check the service manual to see what type of adhesive or sealant you should use. Record the manufacturer's recommendations.

_____

_____

11. Install the vibration damper (harmonic balancer) by carefully pounding on it, or using a special installation tool. In most cases, the damper is installed until it bottoms out against the oil slinger and the timing sprocket. It is best to stand the engine block on end and support the crankshaft if the damper must be pounded on. Check the service manual to see how the damper should be installed. Describe the procedure here.

_____

_____

_____

12. Some vibration dampers are held to the crankshaft by a retaining bolt. Be sure to install the large washer behind the retaining bolt on these engines. Tighten this bolt to specifications. The specifications are:

_____

13. Before installing the valve cover, make sure the cover's sealing flange is flat, and then apply contact adhesive to the valve cover's sealing surfaces in small dabs. Mount the valve cover gasket on the valve cover and align it in position. If the gasket has mounting tabs, use them in tandem with the contact adhesive. Allow the adhesive to dry completely before mounting the valve cover on the cylinder head. Torque the mounting bolts to specifications. The specifications are:

_____

14. The intake manifold gasket seals the joint between the intake manifold and the cylinder head. To be sure the gasket will seal, check its fit before installing it. On steel shim-type gaskets, it is necessary to put a thin and even coat of positioning sealant around the vacuum port openings and a small bead of RTV silicone around the coolant openings. Install the intake manifold with a new gasket and tighten the fastening bolts to the recommended torque specification. The specifications are:

_____

15. Install the thermostat and water outlet housing. Install the thermostat with the temperature sensor facing into the block. Make sure the gasket is positioned properly. Use the sealant recommended by the manufacturer. Take care not to tighten the housing unevenly. Tighten each mounting bolt a little at a time and tighten to specifications. The specifications are:

_____

16. Exhaust manifolds may or may not require a gasket; check the service manual. Also check the manual for tightening sequence and torque specifications. Record this information here.

_____

_____

17. Place the exhaust manifold into position. Tighten the bolts in the center of the manifold first to prevent cracking it. If there are dowel holes in the exhaust manifold that align with dowels in the cylinder head, make sure that these holes are larger than the dowels.   ☐ Task completed

18. Reinstall the engine sling to remove the engine from the engine stand.   ☐ Task completed

19. Raise the engine into the air on a suitable hoist, and remove the engine stand mounting head.   ☐ Task completed

20. Set the assembled engine on the floor and support it with blocks of wood while attaching the flywheel or flex plate.   ☐ Task completed

21. Make sure you use the right flywheel bolts and lock washers. These bolts have very thin heads and the lock washers are thin. Make sure that the bolts are properly torqued and tightened in the correct sequence. The torque specifications are:

_____

_____

22. If the vehicle has a manual transmission, install the clutch. Make sure the transmission's pilot bushing or bearing is in place in the rear of the crankshaft and that it is in good condition. Does the vehicle have a manual transmission? What condition is the plot bearing or bushing in?

_____

_____

23. Assemble the clutch onto the flywheel. Start the bolts by hand. Make sure the disc is installed in the right direction. There should be a marking on it that says "flywheel side."   ☐ Task completed

24. Using a clutch-aligning tool, align the clutch disc. Then, tighten the disc and pressure plate to the flywheel. Tighten the bolts in the proper sequence and to the correct torque. The specifications are:

_____

25. On cars equipped with automatic transmissions, install the torque converter, making sure that it is correctly engaged with the transmission's front pump. The drive lugs on the converter should be felt engaging the transmission front pump gear.   ☐ Task completed

26. The motor mount bolts may now be installed loosely on the block. The bolts are left loose during engine installation so that the mounts can be easily aligned with the front mount brackets. Check the condition of the mounts and describe your findings here.

_____

_____

_____

Instructor's Comments

_____

_____

_____

ENGINE REPAIR JOB SHEET 39

Checking Balance Shafts

Name _____ Station _____ Date _____

NATEF Correlation

This Job Sheet addresses the following NATEF task:

> **C.13.** Inspect auxiliary (balance, intermediate, idler, counterbalance or silencer) shaft(s); inspect shaft(s) and support bearings for damage and wear; determine necessary action; reinstall, and time.

Objective

Upon completion of this job sheet, you will be able to inspect the auxiliary shaft(s) (balance, intermediate, idler, counterbalance or silencer), as well as inspect the support bearings for damage and wear.

Tools and Materials

Micrometer

Telescoping gauge

Service manual

Protective Clothing

Goggles or safety glasses with side shields

Describe the vehicle being worked on:

Year _____ Make _____ Model _____

VIN _____ Engine type and size _____

PROCEDURE

1. What is the purpose of the auxiliary shaft found in the engine you are working on?

   _____

   _____

2. With the shaft removed from the engine, carefully inspect its condition. Pay close attention to bearing journals. Summarize your findings.

   _____

   _____

3. Measure the OD of the bearing journals and compare them to the specifications given in the service manual. Summarize your findings.

   _____

   _____

4. Visually inspect the bearings for the shaft and run your finger over the bearing surface to check for scoring or other damage. Summarize your findings.

_____

_____

5. With a telescoping gauge, measure the ID of the bearings and record your findings.

_____

_____

6. Check the bearings for out-of-roundness and taper by measuring the bearing ID at 90-degree intervals and by measuring the bearings at their center and outside edges. Do your measurements indicate that the bearing is round and not tapered?

_____

_____

7. Subtract the OD or the bearing journals from the ID of the bearings. This is the oil clearance. What is your result?

_____

_____

8. Compare your measured oil clearance to the specs and summarize your results.

_____

_____

9. Based on these checks, what services do you recommend?

_____

_____

Instructor's Comments

_____

_____

_____

ENGINE REPAIR JOB SHEET 40

Prime an Engine's Lubrication System

Name _____ Station _____ Date _____

NATEF Correlation

This Job Sheet is related to the following NATEF task:

C.15. Assemble the engine using gaskets, seals, and formed-in-place (tube-applied) sealants, thread sealers, etc. according to manufacturer's specifications.

Objective

Upon completion of this job sheet, you will be able to prime an engine's lubrication system after the engine has been rebuilt.

Tools and Materials

Reversible electric drill

Priming tool

Basic hand tools

Protective Clothing

Goggles or safety glasses with side shields

Describe the vehicle being worked on:

Year _____ Make _____ Model _____

VIN _____ Engine type and size _____

PROCEDURE

NOTE: *After an engine has been rebuilt, it must be pre-oiled to prevent damage to engine parts when it is first started. This job sheet gives the procedure for manual pre-oiling. Fill out the part of the job sheet that matches the equipment and engine you are working with. Also make sure you follow the guidelines given in the service manual.*

Manual Pre-Oiling

1. Make sure the engine has the proper amount of oil and the right type. Then, remove the distributor. ☐ Task completed

2. If you have a priming tool to drive the oil pump drive shaft, move to the next step. If you do not, select a socket that fits positively on the oil pump drive shaft located in the distributor shaft bore in the engine block. What size socket will you use?

_____

3. Affix the priming tool or socket with an extension to the electric drill. ☐ Task completed

4. Refer to the service manual and determine what direction the distributor normally rotates in. Is it clockwise or counterclockwise?

_____

5. Set the drill to rotate in that direction.   ☐ Task completed

6. Fit the priming tool or socket onto the oil pump drive.   ☐ Task completed

7. Spin the drive until a resistance to turning is felt. What does this indicate?

_____

_____

8. Continue to spin for an additional 30 seconds.   ☐ Task completed

9. Stop spinning the drive and rotate the crankshaft 90 degrees and repeat   ☐ Task completed
   the process.

10. Again stop spinning the drive and rotate the crankshaft 90 degrees and   ☐ Task completed
    repeat the process.

11. And one last time, stop spinning the drive and rotate the crankshaft 90 degrees and repeat the process. Why is it necessary to rotate the crankshaft?

_____

_____

_____

12. Reinstall the distributor, making sure it is positioned properly.   ☐ Task completed

Alternative Manual Pre-Oiling

1. Make sure the oil filter is filled with clean, fresh oil before installing it. Also   ☐ Task completed
   make sure the engine has the proper amount of oil and the right type.

2. Connect the battery terminals and disable the ignition.   ☐ Task completed

3. Crank the engine with the starter for 30 seconds, then stop.   ☐ Task completed

4. Wait at least 30 seconds, then crank the engine again for 30 seconds, then stop. Why do you need to stop after cranking for 30 seconds?

_____

_____

5. Repeat the process above as many times as necessary. The goal is to have an oil pressure reading on the oil pressure gauge or to have the oil warning light turn off. How many times did you need to repeat this process?

_____

_____

6. Reconnect the ignition system and start the engine. Check the oil pressure at idle. What was the pressure? Does it match specifications?

_____

_____

_____

_____

Instructor's Comments

_____

_____

_____

ENGINE REPAIR JOB SHEET 41

Measure Engine Oil Pressure

Name _____ Station _____ Date _____

NATEF Correlation

This Job Sheet addresses the following NATEF task:

D.1. Perform oil pressure tests; determine necessary action

Objective

Upon completion of this job sheet, you will be able to perform oil pressure tests on an engine.

Tools and Materials

Adapter fitting   Open-end wrench

Fender covers   Service manual

Oil-pressure test gauge   Tachometer

Protective Clothing

Safety goggles or glasses with side shields

Describe the vehicle being worked on:

Year _____ Make _____ Model _____

VIN _____ Engine type and size _____

Describe general condition:

_____

PROCEDURE

1. Look up the oil pressure specifications for this engine in the appropriate   ☐ Task completed
   service manual:

   Oil pressure _____

   Engine speed _____

2. Locate the oil-pressure sending unit; usually it is on the lower side of the   ☐ Task completed
   engine block. Disconnect the wire from the sending unit and use an open-
   end wrench to remove the sender.

3. Tighten the oil-pressure test gauge into the hole in the block where the   ☐ Task completed
   sender was removed. Use an adapter fitting, if necessary, to make the
   connection.

4. Check the engine's oil level and fill, if required.   ☐ Task completed

WARNING: *Be extremely careful when working near a running engine. Always wear safety goggles or glasses with side shields when working around moving machinery and be sure that your clothing is not loose.*

5. Start the engine and observe the pressure reading on the gauge. Make sure the engine speed is set to specifications for testing pressure. If necessary, use a tachometer and adjust the engine idle speed. Record the measured oil pressure, then turn the engine off.   ☐ Task completed

_____

6. Is your measurement below specifications?   ☐ Task completed

_____

7. Consult the appropriate service manual and list your conclusions and recommendations.   ☐ Task completed

_____

_____

8. Remove the test gauge and adapter fitting. Reinstall the oil pressure sender and connect the wire. Start the engine and check for leaks.   ☐ Task completed

Problems Encountered

_____

_____

_____

Instructor's Comments

_____

_____

_____

ENGINE REPAIR JOB SHEET 42

Test Cooling System

Name _____ Station _____ Date _____

NATEF Correlation

This Job Sheet addresses the following NATEF tasks:

D.3. Perform cooling system, cap, and recovery system tests (pressure, combustion leakage, and temperature); determine necessary action.

D.7. Test coolant; drain and recover coolant; flush and refill cooling system with recommended coolant; bleed air as required

Objective

Upon completion of this job sheet, you will be able to perform cooling system, cap, and recovery system pressure, combustion leakage, and temperature tests, as well as test, drain, and recover coolant, and flush and refill the cooling system with the recommended coolant and bleed the air as required.

Tools and Materials

Clean cloth rags
Cooling system tester
Coolant hydrometer
Drain pan
Hand tools
Cooling system flusher
Coolant recycler

Protective Clothing

Goggles or safety glasses with side shields

Describe the vehicle being worked on:

Year _____ Make _____ Model _____

VIN _____ Engine type and size _____

Describe general condition:

_____

PROCEDURE

1. Wipe off the radiator filler neck and inspect it.   ☐ Task completed

   Is the sealing seat free of accumulated dirt, nicks, or anything that might   ☐ Yes ☐ No
   prevent a good seal?

Are the cams on the out-turned flange bent or worn? ☐ Yes ☐ No

2. Inspect the tube from the radiator to the expansion tank for dents and other obstructions. Run wire through the tube to be certain it is clear. ☐ Task completed

3. To test the cooling system for external leaks, attach a cooling system tester to the appropriate flexible adapter and carefully pump up pressure. Looking at the tester's gauge, bring the pressure up to the proper cooling system test point indicated on the dial face. ☐ Task completed

Does the pressure begin to drop? ☐ Yes ☐ No

If so, check all external connections, including hoses, gaskets, and heater core for leaks. ☐ Task completed

4. Check all points closely for small pinhole leaks or potentially dangerous weak points in the system. ☐ Task completed

To test for internal leaks:

5. Remove the flexible adapter from radiator, and replace the pressure cap. ☐ Task completed

6. Start the engine and allow it to reach normal operating temperature so the thermostat will open fully. ☐ Task completed

7. Slowly and carefully remove the pressure cap and replace it with the flexible adapter. ☐ Task completed

WARNING: *Engine coolant will be under pressure. Always wear hand protection and safety goggles or glasses with side shields when doing this task.*

8. Lock the tester slowly into place. Watch for pressure buildup. ☐ Task completed

WARNING: *If the pressure builds up suddenly due to an internal leak, remove the tester immediately.* ☐ Yes ☐ No

9. If no immediate pressure buildup is visible on the gauge, keep the engine running. Pressurize the cooling system. ☐ Task completed

10. Does the gauge fluctuate? If so, the cooling system has a compression or combustion leak. ☐ Yes ☐ No

11. Before removing it, hold the tester body and press the pressure release button against some object on the car to relieve the pressure in the system. ☐ Task completed

WARNING: *Always wear hand protection and safety goggles or glasses with side shields when doing this task. Avoid the hot coolant and steam.*

12. After the pressure has been relieved, shield your hands with a cloth wrapped around the flexible adapter and filler neck. Slowly turn the adapter cap from the lock position to the safety unlock position of the radiator filler neck cam. Do not detach the flexible adapter from the filler neck. Allow the pressure to dissipate completely in this safety position. Then you can remove the adapter. ☐ Task completed

13. Summarize your findings from these checks.

_____

_____

14. Draw some coolant out of the radiator or recovery tank with the coolant hydrometer. ☐ Task completed

15. Check the strength of the coolant by observing the position of the tester's float. Summarize what you found out about the coolant and what service you recommend.

_____

_____

16. Reinstall the radiator cap, then start the engine and move the heater control to its full heat position. ☐ Task completed

17. Turn off the engine after it has only slightly warmed up. ☐ Task completed

18. Place a drain pan under the radiator drain plug. ☐ Task completed

19. Make sure the engine and cooling system are not hot, and then open the drain plug. Keep the radiator cap on until the recovery tank is emptied. ☐ Task completed

20. Once all coolant has drained, close the drain plug. Then empty the drain pan with the coolant into the coolant recycler and process the used coolant according to the manufacturer's instructions. ☐ Task completed

21. Connect the flushing machine to the cooling system. ☐ Task completed

22. Follow the manufacturer's instructions for machine operation. Typically, flushing continues until clear water flows out of the system. Once flushing is complete, disconnect the flushing machine. ☐ Task completed

23. Look up the capacity of the cooling system in the service manual. What is it?

_____

24. Prepare to put in a mixture of 50% coolant and 50% water. ☐ Task completed

25. Locate the engine's cooling system bleed valve. Where is it?

_____

26. Loosen the bleed valve. ☐ Task completed

27. Pour an amount of coolant that is equal to half of the cooling system's capacity into the system. ☐ Task completed

28. Add water until some coolant begins to leak from the bleed valve. Then close the valve. ☐ Task completed

29. Leave the radiator cap off and start the engine. ☐ Task completed

30. Continue to add water to the system as the engine warms up. ☐ Task completed

31. Once the system appears full, install the radiator cap. ☐ Task completed

32. Observe the system for leaks and watch the activity in the recovery tank. ☐ Task completed
    If no coolant moves to the tank, the system may need to be bled.

Problems Encountered

_____

_____

_____

Instructor's Comments

_____

_____

_____

ENGINE REPAIR JOB SHEET 43

Inspect, Replace, and Adjust Drive Belts and Pulleys

Name _____ Station _____ Date _____

NATEF Correlation

This Job Sheet addresses the following NATEF task:

D.4. Inspect, replace, and adjust drive belts, tensioners, and pulleys; check pulley and belt alignment.

Objective

Upon completion of this job sheet, you will be able to inspect, replace, and adjust drive belts, tensioners, and pulleys.

Tools and Materials

Hand tools

Belt tension gauge

Pry bar

Service manual

Protective Clothing

Goggles or safety glasses with side shields

Describe the vehicle being worked on:

Year _____ Make _____ Model _____

VIN _____ Engine type and size _____

Describe general condition:

_____

PROCEDURES

1. Visually inspect all drive belts and pulleys for signs of wear, cracks, or breaks.   ☐ Task completed

2. Using a straightedge, check the alignment of the pulleys and describe what you found.

   _____

   _____

3. Loosen all of the mounting brackets on the accessory that is operated by the belt that needs to be replaced.   ☐ Task completed

4. Loosen all brackets on accessories that have belts in front of the belt to be replaced. ☐ Task completed

5. Remove the belt to be replaced. ☐ Task completed

6. Replace the bad belt and reinstall the other belts that were removed for access to the bad belt. ☐ Task completed

7. Use a pry bar to move the accessories and tighten the belts. Be sure the pry bar is not against anything that will bend or break. ☐ Task completed

8. Check the belt tension with the belt tension gauge and adjust to the manufacturer's specifications. Do not overtighten the belts. A belt that is too tight will damage the accessories that are driven by it. ☐ Task completed

9. Tighten all brackets that were loosened to remove the belts. ☐ Task completed

Problems Encountered

_____

_____

_____

Instructor's Comments

_____

_____

_____

ENGINE REPAIR JOB SHEET 44

Servicing Cooling System Hoses

Name _____ Station _____ Date _____

NATEF Correlation

This Job Sheet addresses the following NATEF task:

D.5. Inspect and replace engine cooling and heater system hoses.

Objective

Upon completion of this job sheet, you will be able to inspect and replace cooling and heater system hoses.

Tools and Materials

Basic hand tools

Knife

Wire brush

Emery cloth

Protective Clothing

Goggles or safety glasses with side shields

Describe the vehicle being worked on:

Year _____ Make _____ Model _____

VIN _____ Engine type and size _____

PROCEDURE

1. Carefully check all cooling hoses for leakage, swelling, and chafing. Record your findings here.

   _____

   _____

   _____

2. Squeeze each hose firmly. Do you notice any cracks or signs of splits when the hoses are squeezed? Do any hoses feel mushy or extremely brittle?

   _____

   _____

   _____

3. Carefully examine the areas around the clamps. Are there any rust stains? What is or would be indicated by rust stains?

_____

_____

_____

4. Carefully check the lower radiator hose. This hose contains a coiled wire lining to keep it from collapsing during operation. If the wire loses tension, the hose can partially collapse at high speed and restrict coolant flow. Describe this hose's condition.

_____

_____

5. Identify all hoses that need to be replaced.

_____

_____

_____

_____

_____

6. Gather the new hoses, hose clamps, and coolant.   ☐ Task completed

7. Release the pressure in the system by slowly and carefully opening the radiator cap.   ☐ Task completed

8. Once the pressure is relieved, drain the coolant system below the level that is being worked on. Make sure the drained coolant is collected and recycled or disposed of according to local regulations. What did you do with the drained coolant?

_____

_____

_____

_____

_____

9. Use a knife to cut off the old hose and loosen or cut the old clamp.   ☐ Task completed

10. Slide the old hose off the fitting. If the hose is stuck, do not pry it off; cut it off.   ☐ Task completed

11. Clean off any remaining hose particles with a wire brush or emery cloth. What can happen if dirt or metal burrs are on the fitting when a new hose is attached?

_____

_____

12. Coat the surface with a sealing compound. What compound did you use?

_____

13. Place the new clamps on each end of the hose before positioning the hose.   ☐ Task completed

14. Slide the clamps to about 1/4 inch from the end of the hose after it is properly positioned on the fitting.   ☐ Task completed

15. Tighten the clamp securely. Do not overtighten.   ☐ Task completed

16. Refill the cooling system with coolant.   ☐ Task completed

17. Run the engine until it is warm.   ☐ Task completed

18. Bleed the system according to the manufacturer's recommendations. What did you do to bleed the system?

_____

_____

_____

_____

19. Turn off the engine and recheck the coolant level.   ☐ Task completed

20. Retighten the heater and cooling system hose clamps.   ☐ Task completed

Instructor's Comments

_____

_____

_____

ENGINE REPAIR JOB SHEET 45

Servicing a Thermostat

Name _____ Station _____ Date _____

NATEF Correlation

This Job Sheet addresses the following NATEF task:

D.6. Inspect, test, and replace thermostat and housing.

Objective

Upon completion of this job sheet, you will be able to inspect, test, and replace a thermostat and thermostat housing.

Tools and Materials

Thermometer   Heat source for heating the water

Container for water   Basic hand tools

Protective Clothing

Goggles or safety glasses with side shields

Describe the vehicle being worked on:

Year _____ Make _____ Model _____

VIN _____ Engine type and size _____

PROCEDURE

1. Thoroughly inspect the area around the thermostat and its housing. Describe the result of that inspection.

   _____

   _____

   _____

2. There are several ways to test the opening temperature of a thermostat. The first method does not require that the thermostat be removed from the engine. Begin by removing the radiator pressure cap from a cool radiator and insert a thermometer into the coolant.   ☐ Task completed

3. Start the engine and let it warm up. Watch the thermometer and the surface of the coolant. When the coolant begins to flow or move in the radiator, what is indicated?

   _____

   _____

4. When the fluid begins to flow, record the reading on the thermometer. If the engine is cold and coolant circulates, this indicates the thermostat is stuck open and must be replaced. The measured opening temperature of the thermostat was: _____

5. Compare the measured opening temperature with the specifications. The specifications call for the thermostat to open at what temperature? _____ What do you recommend based on the results of this test?

_____

_____

6. The other way to test a thermostat is to remove it. Begin removal by opening the radiator pressure cap to relieve pressure, if the cap is still on.   ☐ Task completed

7. Drain coolant from the radiator until the level of coolant is below the thermostat housing. Recycle the coolant according to local regulations. What did you do with the drained coolant?

_____

_____

_____

8. Disconnect the upper radiator hose from the thermostat housing.   ☐ Task completed

9. Unbolt and remove the housing from the engine. The thermostat may come off with the housing.   ☐ Task completed

10. Thoroughly clean the gasket surfaces for the housing. Make sure the surfaces are not damaged while doing so and that gasket pieces do not fall into the engine at the thermostat bore.   ☐ Task completed

11. Carefully inspect the thermostat housing. Describe your findings.

_____

_____

12. Suspend the thermostat while it is completely submerged in a small container of water. Make sure it does not touch the bottom of the container. What did you use to suspend it?

_____

_____

_____

13. Place a thermometer in the water so that it does not touch the container and only measures water temperature.   ☐ Task completed

14. Heat the water. When the thermostat valve barely begins to open, read the thermometer. What was the measured opening temperature of the thermostat? Compare that with the specifications.

_____

_____

15. Remove the thermostat from the water and observe the valve. Record what happened and your conclusions about the thermostat.

_____

_____

_____

16. Carefully look over the new (or old if okay) thermostat. Identify which end of the thermostat should face toward the radiator. How did you determine the proper direction for installation?

_____

_____

17. Fit the thermostat in the recessed area in the engine or thermostat housing. Where was the recess? _____

18. Refer to the installation instructions on the gasket's container. Should an adhesive and/or sealant be used with the gasket? If so, what?

_____

_____

19. Install the gasket according to the instructions.   ☐ Task completed

20. Install the thermostat housing. Before tightening it in place, make sure it   ☐ Task completed
    is fully seated and flush onto the engine. Failure to do this will result in a
    broken housing.

21. Tighten the bolts evenly and carefully and to the correct specifications.
    What are the specifications? _____

22. Connect the upper radiator hose to the housing with a new clamp. Why should you use a new clamp?

_____

_____

_____

23. Pressurize the system and check for leaks. Why should you do this now?

_____

_____

24. Replenish the coolant and bring it to its proper level. Install the radiator cap.   ☐ Task completed

25. Run the engine until it is at normal operating temperature. Did the cooling system warm up properly and does it seem that the thermostat is working properly?

_____

_____

26. Recheck the coolant level.   ☐ Task completed

Instructor's Comments

ENGINE REPAIR JOB SHEET 46

Servicing a Water Pump

Name _____ Station _____ Date _____

NATEF Correlation

This Job Sheet addresses the following NATEF task:

D.8. Inspect, test, remove, and replace water pump.

Objective

Upon completion of this job sheet, you will be able to inspect, test, remove, and replace a water pump.

Tools and Materials

Stethoscope

Basic hand tools

Protective Clothing

Goggles or safety glasses with side shields

Describe the vehicle being worked on:

Year _____ Make _____ Model _____

VIN _____ Engine type and size _____

PROCEDURE

1. Begin your inspection of the water pump by looking for evidence of leaks. Look carefully for signs of leakage at the weep hole in the water pump casting. If there is a leak at the weep hole, what is indicated? What did you find during your general inspection of the water pump?

   _____

   _____

2. If you suspect that the water pump is related to a noise problem or if it appears that the water pump seal is leaking, check for the following. Record your findings next to each item.

   a. A bent fan.

   _____

   b. A piece of the fan is missing.

   _____

   c. A cracked fan blade.

   _____

d. Fan mounting surfaces that are not clean or flush.

_____

e. A worn fan clutch.

_____

3. To check the water pump, start the engine and listen for a bad bearing, using a mechanic's stethoscope. Place the stethoscope on the bearing or pump shaft. Describe what you heard and what this indicates.

_____

_____

_____

4. Turn the engine off and remove the fan belt and shroud. Grasp the fan and attempt to move it in and out and up and down. How much were you able to move it and what is indicated by this?

_____

_____

5. Reinstall the fan belt and shroud. Make sure the belt is tightened to the correct tension.   □ Task completed

6. Now, warm up the engine and run it at idle speed. Squeeze the upper hose connection with one hand and accelerate the engine with the other hand. What do you feel and why do you feel it?

_____

_____

7. If the water pump is defective or if it leaks, it must be replaced. Begin replacement by draining the coolant from the cooling system.   □ Task completed

8. Make sure you catch all the coolant as it drains and recycle it as directed by local regulations.   □ Task completed

9. Remove all components, such as the drive belts, fan, fan shroud, shaft spacers, or viscous drive clutch, that block accessibility to the water pump. What do you need to remove in order to remove the water pump?

_____

_____

_____

10. Loosen and remove the bolts in a crisscross pattern from the center outward. Insert a rag into the block opening and scrape off any remains of the old gasket.   □ Task completed

11. When replacing a water pump, always follow the procedures recommended by the manufacturer. What are those recommendations?

_____

_____

_____

12. What should be done to the gasket and sealing surfaces prior to installation of the water pump?

_____

_____

13. Install the mounting bolts and tighten them evenly in a staggered sequence to the torque specifications. What are those specifications?

_____

14. Check the pump to make sure it rotates freely. What happens when you rotate the pump?

_____

_____

15. Reinstall everything you had to remove to gain access to the water pump. What components did you need to install?

_____

_____

_____

16. Make sure the drive belt(s) are tightened to the proper tension. What is the required tension for each belt?

_____

17. Refill the cooling system to the proper level.   ☐ Task completed

18. Pressure check the system and look for signs of leakage.   ☐ Task completed

19. If no leaks are evident, start the engine and allow it to run until it reaches normal operating temperature. Then bleed the air from the system. How did you do that?

_____

_____

20. Shut the engine off and look for leaks, then replenish the cooling system if   ☐ Task completed
    necessary.

Instructor's Comments

ENGINE REPAIR JOB SHEET 47

Remove and Replace a Radiator

Name _____ Station _____ Date _____

NATEF Correlation

This Job Sheet addresses the following NATEF task:

D.9. Remove and replace radiator.

Objective

Upon completion of this job sheet, you will be able to remove and replace a radiator.

Tools and Materials

Drain pan

Protective Clothing

Goggles or safety glasses with side shields

Describe the vehicle being worked on:

Year _____ Make _____ Model _____

VIN _____ Engine type and size _____

PROCEDURE

1. Disconnect the negative terminal of the battery.   □ Task completed

2. Drain the cooling system and recycle or dispose of the coolant according   □ Task completed
   to local regulations.

3. Loosen and remove the hose clamps for the hoses that connect to the   □ Task completed
   radiator.

4. Disconnect the transmission cooler (if so equipped) lines and plug them.   □ Task completed
   What did you use to plug them?

   _____

5. Disconnect the wiring harness connector to the electric cooling fan, if so   □ Task completed
   equipped.

6. Disconnect any sensor wires that may be attached to the radiator. What
   sensors are installed in the radiator?

   _____

   _____

7. Remove the fasteners that attach the cooling fan assembly to the radiator. Or remove the attaching bolts for the fan shrouding. Which did you need to remove?

   _____

   _____

8. Remove the cooling fan assembly or shroud. What else did you need to remove?

   _____

   _____

9. Check the service manual to see if the air conditioning condenser must be removed with the radiator. If so, what must you do to disconnect the condenser from the A/C system?

   _____

   _____

   _____

   _____

10. Remove the upper radiator mounts.   ☐ Task completed

11. Remove the radiator.   ☐ Task completed

12. Place the new or rebuilt radiator into its lower mounts. Make sure the rubber insulators are in place after doing this.   ☐ Task completed

13. Install and bolt the upper mounts.   ☐ Task completed

14. If the A/C condenser was disconnected, reconnect it according to the manufacturer's recommendations.   ☐ Task completed

15. Install all parts of the cooling fan assembly or shroud.   ☐ Task completed

16. Connect any sensor wires that were disconnected during the removal of the radiator.   ☐ Task completed

17. Connect the electrical connector for the cooling fan.   ☐ Task completed

18. Unplug the transmission cooler fittings and attach the lines to the radiator.   ☐ Task completed

19. Inspect the upper and lower radiator hoses. If they are in good shape, reuse them. If they are questionable or bad, replace them. Describe the condition of the hoses.

   _____

   _____

20. With new clamps, attach the upper and lower radiator hoses to the radiator.   ☐ Task completed

21. Refill the cooling system with coolant. How much coolant did you need to add?

_____

_____

22. Pressure check the system and look for evidence of leaks. Describe the results of this check.

_____

_____

23. If there are no leaks, connect the battery.   ☐ Task completed

24. Start the engine and look for leaks.   ☐ Task completed

25. Allow the engine to reach normal operating temperature, then bleed any   ☐ Task completed
    air that was trapped in the cooling system.

26. Check and replenish the transmission fluid level.   ☐ Task completed

27. Turn off the engine and recheck the coolant level.   ☐ Task completed

Instructor's Comments

_____

_____

_____

ENGINE REPAIR JOB SHEET 48

Clean, Inspect, Test, and Replace Electric Cooling Fans and Cooling System-Related Temperature Sensors

Name _____ Station _____ Date _____

NATEF Correlation

This Job Sheet addresses the following NATEF task:

D.10. Inspect, and test fan(s) (electrical or mechanical), fan clutch, fan shroud, and air dams.

Objective

Upon completion of this job sheet, you will be able to inspect and test electrical or mechanical cooling fan(s), a fan clutch, fan shroud, and air dams.

Tools and Materials

Hand tools

Test light (circuit tester)

DMM

Jumper wire

Service manual

Protective Clothing

Goggles or safety glasses with side shields

Describe the vehicle being worked on:

Year _____ Make _____ Model _____

VIN _____ Engine type and size _____

Describe general condition:

_____

PROCEDURE

1. Visually inspect the electric cooling fan(s) and related wiring to determine if they are clean and properly connected. ☐ Task completed

2. Run the engine until it reaches normal operating temperature to determine if the electric cooling fan is working. ☐ Task completed

3. If the fan is operating properly, blow the fan and surrounding area clean with air. ☐ Task completed

4. If the fan is not working, check the power wire to the fan to determine if there is power. ☐ Task completed

5. Check the ground circuit to determine if the system has a good ground. □ Task completed

6. If power and a good ground are present, remove the electric fan assembly and replace it. □ Task completed

7. If there is no power, check the fuse and fan relay. The location of the fuse and relay can be found in the service manual. □ Task completed

8. Replace bad components, then retest fan operation. □ Task completed

9. If there is not a good ground, check the temperature sensor that controls the electric fan. The location of the sensor and the proper test procedure can be found in the service manual. □ Task completed

10. If the sensor fails the test, replace it. Check the operation of the electric fan. □ Task completed

   WARNING: *The electric cooling fans can operate with the engine and the key both turned off. The fan can come on at any time if the engine is hot or there is a problem with a sensor. Keep your hands away from the fan at all times.*

Problems Encountered

_____

_____

_____

Instructor's Comments

_____

_____

_____

ENGINE REPAIR JOB SHEET 49

Service Auxiliary Oil Coolers

Name _____ Station _____ Date _____

NATEF Correlation

This Job Sheet addresses the following NATEF task:

D.11. Inspect auxiliary oil coolers; determine necessary action.

Objective

Upon completion of this job sheet, you will be able to inspect and replace auxiliary oil coolers.

Tools and Materials

OSHA-approved air nozzle

Bucket of water

Vacuum pump

Drain pan

Line plugs

Clean rags

Hand tools

Protective Clothing

Goggles or safety glasses with side shields

Describe the vehicle being worked on:

Year _____ Make _____ Model _____

VIN _____ Engine type and size _____

PROCEDURE

1. Where is the auxiliary oil cooler located?

   _____

   _____

2. Carefully check the lines connecting the cooler to the engine. Are there signs of leakage or damage? Summarize your findings.

   _____

   _____

3. Carefully check the cooler. Are there signs of leakage or damage? Summarize your findings.

_____

_____

4. Place a drain pan under the cooler.   ☐ Task completed

5. Disconnect the cooler lines and remove the cooler.   ☐ Task completed

6. To leak test the cooler, a vacuum pump or shop air can be used. Using a line plug, close off the outlet fitting of the cooler.   ☐ Task completed

7. Connect the vacuum pump line to the inlet fitting at the cooler.   ☐ Task completed

8. Run the vacuum pump. If there is a leak in the cooler, vacuum will not build. What did you observe?

_____

_____

9. Insert the cooler into a bucket of water, making sure water does not enter into it.   ☐ Task completed

10. Insert the air nozzle into the inlet fitting at the cooler.   ☐ Task completed

11. Release air pressure into the cooler while observing the water in the bucket. If there are bubbles, the cooler leaks. What did you observe?

_____

_____

12. If the oil cooler checks out fine, remount it to the vehicle.   ☐ Task completed

13. Using new O-rings and/or seals, connect the lines to the cooler.   ☐ Task completed

14. Look up the capacity of the cooler and add that much oil to the engine.   ☐ Task completed

15. Start the engine and check for leaks.   ☐ Task completed

16. After the engine is warmed up, turn off the engine and recheck the engine's oil level.   ☐ Task completed

Instructor's Comments

_____

_____

_____

ENGINE REPAIR JOB SHEET 50

Servicing Oil Pressure and Temperature Sensors

Name _____ Station _____ Date _____

NATEF Correlation

This Job Sheet addresses the following NATEF task:

D.12. Inspect, test, and replace oil temperature and pressure switches and sensors.

Objective

Upon completion of this job sheet, you will be able to inspect, test, and replace oil temperature and pressure switches and sensors.

Tools and Materials

Ohmmeter

Basic hand tools

Oil pressure tester

Protective Clothing

Goggles or safety glasses with side shields

Describe the vehicle being worked on:

Year _____ Make _____ Model _____

VIN _____ Engine type and size _____

PROCEDURE

1. Describe the type(s) of oil gauges and/or warning lights the engine is equipped with.

   _____

   _____

   _____

2. Explain the purpose of each.

   _____

   _____

   _____

3. Locate the specifications for each in the service manual and record the specifications here.

_____

_____

_____

4. When you turn the ignition on with the engine off, do the indicator lamps light? What are the readings on the gauges?

_____

_____

5. Start the engine. Do the lamps turn off? Do the readings on the gauges change?

_____

_____

6. What can you conclude so far?

_____

_____

_____

7. With the engine off, carefully examine the area around each of the sensors. Look for signs of oil leakage and record your findings.

_____

_____

8. Also look for oil inside the protective boots for the electrical connectors. What can you conclude?

_____

_____

9. Disconnect the electrical connector to each of the sensors and connect an ohmmeter from the terminal of the sensor to ground. Compare the reading to the specifications. What can you conclude from this test?

_____

_____

10. Start the engine and look at the ohmmeter reading. Compare the reading to the specifications. What can you conclude from this test?

_____

_____

11. If the oil pressure sensor is suspected of being bad, note the oil pressure reading shown on the oil pressure gauge on the instrument panel while the engine is idling.

_____

12. Then turn the engine off and remove the sensor. Connect an oil pressure tester to the sensor's bore.   ☐ Task completed

13. Start the engine and allow it to idle. Record the oil pressure shown on the test gauge.

_____

14. Compare the reading on the test gauge with the reading taken at the instrument panel. What can you conclude?

_____

_____

15. Install the sensor or a new one and reconnect the electrical connector to it.   ☐ Task completed

16. If the sensor(s) tested fine but there is still misinformation at the gauges or indicator lamps, the gauge or lamp circuit must be tested. Is it necessary to do so on this vehicle?

_____

_____

_____

_____

Instructor's Comments

_____

_____

_____

ENGINE REPAIR JOB SHEET 51

Perform Oil and Filter Change

Name _____ Station _____ Date _____

NATEF Correlation

This Job Sheet addresses the following NATEF task:

D.13. Perform oil and filter change.

Objective

Upon completion of this job sheet, you will be able to properly change an engine's oil and oil filter.

Tools and Materials

Rags

Funnel

Oil filter wrench

Lift

Protective Clothing

Goggles or safety glasses with side shields

Describe the vehicle being worked on:

Year _____ Make _____ Model _____

VIN _____ Engine type and size _____

PROCEDURE

1. Always make sure the vehicle is positioned safely on a lift or supported by   ☐ Task completed
   jack stands. Before raising the vehicle, allow the engine to run awhile. After
   it is warm, turn off the engine.

2. Place the oil drain pan under the drain plug before beginning to drain the   ☐ Task completed
   oil.

3. Loosen the drain plug with the appropriate wrench. After the drain plug
   is loosened, quickly remove it so the oil can freely drain from the oil pan.
   Make sure the drain pan is positioned so it can catch all of the oil. Describe
   the color and condition of the oil.

   _____

   _____

4. While the oil is draining, use an oil filter wrench to loosen and remove the oil filter. Describe what you needed to do in order to do this.

_____

_____

5. Make sure the oil filter seal came off with the filter. Then place the filter into the drain pan so it can drain. After it has completely drained, discard the filter according to local regulations. What are your local requirements for disposing of oil filters and oil?

_____

_____

6. Wipe off the oil filter sealing area on the engine block. Then apply a coat of clean engine oil onto the new filter's seal.   ☐ Task completed

7. Install the new filter and hand-tighten it. Oil filters should be tightened according to the directions given on the filter. What are the instructions?

_____

_____

8. Prior to installing the drain plug, wipe off its threads and sealing surface with a clean rag. Inspect the threads and describe their condition:

_____

_____

9. The drain plug should be tightened according to the manufacturer's recommendations. Overtightening can cause thread damage, while undertightening can cause an oil leak. What is the tightening spec for the plug?

_____

10. With the oil filter and drain plug installed, lower the vehicle and remove   ☐ Task completed
the oil filler cap.

11. Carefully pour the oil into the engine. The use of a funnel usually keeps oil from spilling onto the engine. What type of oil is recommended and how much?

_____

12. After the recommended amount of oil has been put into the engine, check   ☐ Task completed
the oil level.

13. Start the engine and allow it to reach normal operating temperature. While the engine is running, check the engine for oil leaks, especially around the oil filter and drain plug. If there is a leak, shut down the engine and correct the problem. Were there any leaks? Where?

_____

_____

14. After the engine has been turned off, recheck the oil level and correct it as   ☐ Task completed
necessary.

Instructor's Comments

NOTICE: SOME PARTS OF THIS COPY ARE DIFFERENT THAN THE PREVIOUS PRINTING OF THIS BOOK.

The contents of this book have been updated in response to the recent changes made by NATEF. Most of these changes were minor and involved the rewording of task statements. There were, however, some new tasks added to their list. The additions resulted in the renumbering of the tasks, those changes have also been made to this workbook.

The NATEF task list included here shows where the additions, deletions, and changes were made. There is also a table that shows which Job Sheet correlates to the new NATEF task numbers, as well as the old.

Job Sheets that relate to the new tasks follow the job sheet correlation chart.

NATEF TASK LIST FOR ENGINE REPAIR

> *Legend:*   everything that is **new** is <u>underlined</u>
> everything that has been **deleted** is ~~struck through~~

A. General Engine Diagnosis; Removal and Reinstallation (R&R)

A.1.	~~Verify~~ <u>Identify</u> and interpret engine concern; determine necessary action.	Priority Rating 1
<u>A.2.</u>	<u>Research applicable vehicle and service information, such as internal engine operation, vehicle service history, service precautions, and technical service bulletins.</u>	
<u>A.3.</u>	<u>Locate and interpret vehicle and major component identification numbers (VIN, vehicle certification labels, and calibration decals).</u>	<u>Priority Rating 1</u>
A.~~2~~<u>4</u>.	Inspect engine assembly for fuel, oil, coolant, and other leaks; determine necessary action.	Priority Rating 2
A.~~3~~ <u>5.</u>	Diagnose engine noises and vibrations; determine necessary action.	Priority Rating 3
A.~~4~~ <u>6.</u>	Diagnose the cause of excessive oil consumption, unusual engine exhaust color, odor, and sound; determine necessary action.	Priority Rating 3
A.~~5~~ <u>7.</u>	Perform engine vacuum tests; determine necessary action.	Priority Rating 1
A.~~6~~ <u>8.</u>	Perform cylinder power balance tests; determine necessary action.	Priority Rating 1
A.~~7~~ <u>9.</u>	Perform cylinder compression tests; determine necessary action.	Priority Rating 1
A.~~8~~ <u>10.</u>	Perform cylinder leakage tests; determine necessary action.	Priority Rating 1
A.9.	~~Remove engine (front-wheel-drive); prepare for disassembly.~~	~~Priority Rating 3~~
A.10.	~~Reinstall engine (front-wheel-drive).~~	~~Priority Rating 3~~
A.11.	~~Remove engine (rear-wheel-drive); prepare for disassembly.~~	~~Priority Rating 3~~
A.12.	~~Reinstall engine (rear-wheel-drive).~~	~~Priority Rating 3~~
<u>A.11.</u>	<u>Remove and reinstall engine in a late model front-wheel-drive vehicle (OBD-I or newer); reconnect all attaching components and restore the vehicle to running condition.</u>	<u>Priority Rating 3</u>
<u>A.12.</u>	<u>Remove and reinstall engine in a late model rear-wheel-drive vehicle (OBD-I or newer); reconnect all attaching components and restore the vehicle to running condition.</u>	<u>Priority Rating 3</u>

B. Cylinder Head and Valve Train Diagnosis and Repair

B.1.   Remove cylinder head(s); visually inspect cylinder head(s) for cracks; check gasket surface areas for warpage and leakage; check passage condition.   Priority Rating 2

B.2.   Install cylinder heads and gaskets; tighten according to manufacturer's specifications and procedures.   Priority Rating 2

B.3.   Inspect and test valve springs for squareness, pressure, and free height comparison; ~~replace as needed~~ determine necessary action.   Priority Rating 3

~~B.4.~~   ~~Inspect valve spring retainers, locks, and valve grooves.~~   ~~Priority Rating 2~~

~~B.5.~~   ~~Replace valve stem seals.~~   ~~Priority Rating 3~~

B.4.   Replace valve stem seals on an assembled engine; inspect valve retainers, locks, and valve grooves; determine necessary action.   Priority Rating 2

B.~~6.~~ 5.   Inspect valve guides for wear; check valve guide height and stem-to-guide clearance; ~~recondition or replace as needed.~~ determine necessary action.   Priority Rating 3

B.6.   Inspect valves and valve seats; determine necessary action.

~~B.7.~~   ~~Resurface valves; perform necessary action.~~   ~~Priority Rating 2~~

~~B.8.~~   ~~Resurface valve seats; perform necessary action.~~   ~~Priority Rating 2~~

B.~~9.~~ 7.   Check valve face-to-seat contact and valve seat concentricity (runout); ~~service seats and valves as needed.~~ determine necessary action.   Priority Rating 3

B.~~10.~~ 8.   Check valve spring assembled height and valve stem height; ~~service valve and spring assemblies as needed.~~ determine necessary action.   Priority Rating 2

B.~~11.~~ 9.   Inspect pushrods, rocker arms, rocker arm pivots and shafts for wear, bending, cracks, looseness, and blocked oil passages (orifices); ~~perform necessary action.~~ determine necessary action.   Priority Rating 2

B.~~12.~~10.   Inspect hydraulic or mechanical lifters; ~~replace as needed.~~ determine necessary action.   Priority Rating 2

B.~~13.~~ 11.   Adjust valves (mechanical or hydraulic lifters).   Priority Rating 1

B.~~14.~~ 12.   Inspect camshaft drives (including gear wear and backlash, sprocket and chain wear); ~~replace as necessary.~~ determine necessary action.   Priority Rating 2

B.~~15.~~ 13.   Inspect and replace timing belts(s), overhead camdrive sprockets, and tensioners; check belt/chain tension; adjust as necessary.   Priority Rating 1

B.~~16.~~14.   Inspect camshaft for runout, journal wear and lobe wear.   Priority Rating 3

B.~~17.~~15.   Inspect ~~and measure~~ camshaft bearings surface for wear, damage, out-of-round, and alignment; determine necessary action.   Priority Rating 3

B.~~18.~~16.   ~~Verify~~ Establish camshaft(s) timing and cam sensor indexing according to manufacturer's specifications and procedure.   Priority Rating 1

C. Engine Block Assembly Diagnosis and Repair

~~C.1.~~   ~~Inspect and replace pans, covers, gaskets, and seals.~~   ~~Priority Rating 2~~

C.1.   Disassemble engine block; clean and prepare components for inspection and reassembly.

C.2.   Inspect engine block for visible cracks, passage condition, core and gallery plug condition, and surface warpage; determine necessary action.   Priority Rating 2

C.3.	Inspect internal and external threads; restore as needed (includes installing thread inserts).	Priority Rating 1
~~C.4.~~	~~Remove cylinder wall ridges.~~	~~Priority Rating 3~~
C.~~5.~~ 4.	Inspect and measure cylinder walls for damage and wear; determine necessary action.	Priority Rating 2
C.~~6.~~ 5.	Deglaze and clean cylinder walls.	Priority Rating 1
C.~~7.~~ 6.	Inspect and measure camshaft bearings for wear, damage, out-of-round, and alignment; determine necessary action.	Priority Rating 3
C.~~8.~~ 7.	Inspect crankshaft for <u>end play, straightness, journal damage, keyway damage, thrust flange and sealing surface condition, and visual</u> surface cracks ~~and journal damage~~; check oil passage condition; measure journal wear; <u>check crankshaft sensor reluctor ring (where applicable)</u> ~~determine necessary action~~.	Priority Rating 3
C.~~9.~~ 8.	Inspect and measure main and connecting rod bearings for damage, clearance, and end play; determine necessary action (includes the proper selection of bearings).	Priority Rating 2
C.~~10.~~ 9.	Identify piston and bearing wear patterns that indicate connecting rod alignment and main bearing bore problems; inspect rod alignment and bearing bore condition.	Priority Rating 3
C.~~11.~~ 10.	Inspect <u>and</u>~~,~~ measure~~, and service~~ pistons ~~and pins~~; determine necessary action.	Priority Rating 2
<u>C.11.</u>	<u>Remove and replace piston pins.</u>	<u>Priority Rating 2</u>
C.12.	Inspect, measure, and install piston rings.	Priority Rating 2
C.~~15.~~<u>13.</u>	Inspect auxiliary (balance, intermediate, idler, counter-balance or silencer) shaft(s); inspect shaft(s) and support bearings for damage and wear; determine necessary action; reinstall and time.	Priority Rating 3
C.~~13.~~<u>14.</u>	Inspect, repair or replace crankshaft vibration damper (harmonic balancer).	Priority Rating 3
C.~~14.~~ 15.	~~Re~~Assemble <u>the</u> engine ~~components~~ using ~~correct~~ gaskets<u>, seals, and formed-in-place (tube-applied)</u> ~~and~~ sealants<u>, thread sealers, etc. according to manufacturer's specifications</u>.	Priority Rating 2
~~C.16.~~	~~Prime engine lubrication system.~~	~~Priority Rating 1~~

D. Lubrication and Cooling Systems Diagnosis and Repair

D.1.	Perform oil pressure tests; determine necessary action.	Priority Rating 1
D.2.	Inspect oil pump gears or rotors, housing, pressure relief devices, and pump drive; perform necessary action.	Priority Rating 3
D.3.	Perform cooling system, cap, and recovery system tests (pressure, combustion leakage, and temperature); determine necessary action.	Priority Rating 1
D.4.	Inspect, replace, and adjust drive belts, tensioners, and pulleys<u>; check pulley and belt alignment</u>.	Priority Rating 1
D.5.	Inspect and replace engine cooling and heater system hoses.	Priority Rating 2
D.6.	Inspect, test, and replace thermostat and housing.	Priority Rating 2
D.7.	Test coolant; drain and recover coolant; flush and refill cooling system with recommended coolant; bleed air as required.	Priority Rating 1
D.8.	Inspect, test, remove, and replace water pump.	Priority Rating 2
D.9.	Remove and replace radiator.	Priority Rating 2

D.10.   Inspect, and test fans(s) (electrical or mechanical), fan
clutch, fan shroud, and air dams.   Priority Rating 2

D.11.   Inspect auxiliary oil coolers; ~~replace as needed~~ <u>determine
necessary action.</u>   Priority Rating 3

D.12.   Inspect, test, and replace oil temperature and pressure
switches and sensors.   Priority Rating 2

D.13.   Perform oil and filter change.   Priority Rating 1

New Task #	Old Task #	Job Sheet #
A.1	A.1	1
A.2	NEW	52
A.3	NEW	52
A.4	A.2	1
A.5	A.3	1
A.6	A.4	1
A.7	A.5	2
A.8	A.6	3
A.9	A.7	4
A.10	A.8	5
A.11	NEW	6&7
A.12	NEW	8&9
B.1	B.1	10
B.2	B.2	11
B.3	B.3	12
B.4	NEW	13&14
B.5	B.6	10
B.6	NEW	13 (15 &16 are related)
B.7	B.9	17
B.8	B.10	18
B.9	B.11	19
B.10	B.12	19
B.11	B.13	20
B.12	B.14	21
B.13	B.15	22
B.14	B.16	23
B.15	B.17	24
B.16	B.18	25
C.1	NEW	53 (26 is related)
C.2	C.2	27
C.3	C.3	28
C.4	C.5	30 (29 is related)
C.5	C.6	31
C.6	C.7	32
C.7	C.8	33
C.8	C.9	34
C.9	C.10	35
C.10	C.11	36
C.11	NEW	36
C.12	C.12	37
C.13	C.15	39
C.14	C.13	37
C.15	C.14	38 (40 is related)
D.1	D.1	41
D.2	D.2	26
D.3	D.3	42
D.4	D.4	43
D.5	D.5	44
D.6	D.6	45
D.7	D.7	42
D.8	D.8	46
D.9	D.9	47
D.10	D.10	48
D.11	D.11	49
D.12	D.12	50
D.13	D.13	51

ENGINE REPAIR JOB SHEET 52

Gathering Vehicle Information

Name _____ Station _____ Date _____

NATEF Correlation

This Job Sheet addresses the following NATEF tasks:

A.2. Research applicable vehicle and service information, such as internal engine operation, vehicle service history, service precautions, and technical service bulletins.

A.3. Locate and interpret vehicle and major component identification numbers (VIN, vehicle certification labels, calibration labels).

Objective

Upon completion of this job sheet, you will be able to gather service information about a vehicle and its engine and related systems.

Tools and Materials

Appropriate service manuals

Computer

Protective Clothing

Goggles or safety glasses with side shields

Describe the vehicle being worked on:

Year _____ Make _____ Model _____

VIN _____

PROCEDURE

1. Using the service manual or other information source, describe what each letter and number in the VIN for this vehicle represents.

   _____

   _____

2. Locate the Vehicle Emissions Control Information (VECI) label and describe where you found it.

   _____

   _____

3. Summarize what information you found on the VECI label.

   _____

   _____

4. While looking in the engine compartment did you find a label regarding the specifications for the engine? Describe where you found it.

_____

_____

_____

_____

5. Summarize the information contained on this label.

_____

_____

_____

_____

6. Using a service manual or electronic database, locate the information about the vehicle's engine. List the major components of the system and describe the primary characteristics of the engine (number of valves, shape/configuration, firing order. etc.).

_____

_____

_____

_____

_____

7. Using a service manual or electronic database, locate and record all service precautions regarding working on the engine and its systems as noted by the manufacturer.

_____

_____

_____

_____

8. Using the information that is available, locate and record the vehicle's service history.

_____

_____

_____

_____

9. Using the information sources that are available, summarize all Technical Service Bulletins for this vehicle that relate to the engine and its systems.

_____

_____

_____

Instructor's Comments

ENGINE REPAIR JOB SHEET 53

Disassemble and Clean an Engine Block

Name _____ Station _____ Date _____

NATEF Correlation

This Job Sheet addresses the following NATEF task:

C.1. Disassemble engine block; clean and prepare components for inspection and reassembly.

Objective

Upon completion of this job sheet, you will be able to disassemble an engine block, clean, and inspect the components and prepare all parts for reassembly.

Protective Clothing

Goggles or safety glasses with side shields

Tools and Materials

Appropriate service manuals   Various pullers

Hand tools   Crack detector

Cleaning solution/solvent   Rod bolt protectors

Ridge reamer

Describe the vehicle being worked on:

Year _____ Make _____ Model _____

VIN _____ Engine type and size _____

PROCEDURE

1. Before engine disassembly, be sure the engine is securely bolted to an   ☐ Task completed
   engine stand or sitting on blocks.

2. Look up disassembly instructions in the service manual for the specific
   model car and engine prior to beginning to disassemble the engine.
   Summarize those instructions.

   _____

   _____

3. Remove the valve cover or covers and disassemble the rocker arm
   components. After removing the rocker arm and pushrods, check the
   rocker area for sludge. Excessive buildup can indicate a poor oil
   change schedule and is a signal to look for similar wear patterns on
   other components. Record what you found.

   _____

   _____

4. Remove the pushrods and rocker arms or rocker arm assemblies and keep them in the in exact order.   ☐ Task completed

5. Carefully check the lifters for a dished bottom or scratches. What did you find?

   _____

   _____

6. Loosen the cylinder head bolts one or two turns each, working from the center of the cylinder head outward. Then, remove the bolts, again following the center-outward sequence.   ☐ Task completed

7. Lift the cylinder head off. Save the cylinder head so it can be compared to the new head gasket during reassembly.   ☐ Task completed

8. On overhead cam engines, the camshaft must be removed before the cylinder head can be disassembled. Before tearing down the cam follower assembly, draw a diagram and use a felt-tipped marker to label the parts.   ☐ Task completed

9. On overhead valve engines, remove the timing cover.   ☐ Task completed

10. Remove the harmonic balancer or vibration damper using the puller designed for the purpose.   ☐ Task completed

11. Remove the camshaft. Support the camshaft during removal to avoid dragging lobes over bearing surfaces, which would damage bearings and lobes. Do not bump cam lobe edges, which can cause chipping.   ☐ Task completed

12. After the camshaft has been removed, visually examine it for any obvious defects—rounded lobes, edge wear, galling, and the like. Describe the condition of the camshaft here.

    _____

    _____

13. Remove the oil pan if it was not removed previously. Inspect the flange for distortion and straighten as necessary. What did you find?

    _____

    _____

14. Remove the oil pump.   ☐ Task completed

15. Remove the valve (camshaft) timing components.   ☐ Task completed

16. Inspect the sprockets for wear, cracks, and broken teeth. Inspect the timing gears for excessive backlash. Check the chain for slackness and wear. Record your findings.

    _____

    _____

17. Carefully remove the cylinder ridge with a ridge removing tool. How much of a ridge was there?

    _____

    _____

18. After the ridge removing operation, wipe all the metal cuttings out of the cylinder. Use an oily rag to wipe the cylinder. The cuttings will stick to it.   ☐ Task completed

19. Check all connecting rods and main bearing caps for correct position and numbering. If the numbers are not visible, use a center punch or number stamp to number them. Describe what you found.

_____

_____

20. Position the crankshaft throw at the bottom of its stroke.   ☐ Task completed

21. Remove the connecting rod nuts and cap. Tap the cap lightly with a soft hammer or wood block to aid in cap removal. Cover the rod bolts with protectors to avoid damage to the crankshaft journals.

22. Carefully push the piston and rod assembly out with the wooden hammer handle or wooden drift and support the piston by hand as it comes out of the cylinder.   ☐ Task completed

23. With bearing inserts in the rod and cap, replace the cap (numbers on same side) and install the nuts. Repeat the procedure for all other piston and rod assemblies.   ☐ Task completed

24. Remove the flywheel or flex plate. Mark the position of the crankshaft and flywheel or flexplate.   ☐ Task completed

25. Remove the main bearing cap bolts and main bearing caps.   ☐ Task completed

26. Carefully take out the crankshaft by lifting both ends equally to avoid bending and damage.   ☐ Task completed

27. Remove the main bearings and rear main oil seal from the block and the main bearing caps.   ☐ Task completed

28. Examine the bearing inserts for signs of abnormal engine conditions such as embedded metal particles, lack of lubrication, antifreeze contamination, oil dilution, uneven wear, and wrong or undersized bearings. Record your findings.

_____

_____

29. Carefully inspect the main journals on the crankshaft for damage.   ☐ Task completed

30. Remove all core plugs and oil gallery plugs. Describe what you needed to do to remove them.

_____

_____

31. Visually check the cylinder head and block and their parts for cracks or other damage before they are cleaned.   ☐ Task completed

32. While inspecting the parts, check to see if the engine has ever been torn down before by looking at the bearings for undersized or date codes. What did you find?

_____

_____

33. After the block or cylinder head parts have been removed and disassembled, thoroughly clean them.   ☐ Task completed

34. Describe the type of dirt and the method you used to clean the parts.

_____

_____

_____

_____

Instructor's Comments

_____

_____

_____